Birkhäuser

Oberwolfach Seminars

Volume 51

The workshops organized by the *Mathematisches Forschungsinstitut Oberwolfach* are intended to introduce students and young mathematicians to current fields of research. By means of these well-organized seminars, also scientists from other fields will be introduced to new mathematical ideas. The publication of these workshops in the series *Oberwolfach Seminars* (formerly *DMV seminar*) makes the material available to an even larger audience.

Eberhard Bänsch • Klaus Deckelnick •
Harald Garcke • Paola Pozzi

Interfaces: Modeling, Analysis, Numerics

Birkhäuser

Eberhard Bänsch
Department Mathematik
Universität Erlangen-Nürnberg
Erlangen, Germany

Harald Garcke
Fakultät für Mathematik
Universität Regensburg
Regensburg, Bayern, Germany

Klaus Deckelnick
Institut für Analysis und Numerik
Universität Magdeburg
Magdeburg, Germany

Paola Pozzi
Fakultät für Mathematik
Universität Duisburg-Essen
Essen, Germany

ISSN 1661-237X ISSN 2296-5041 (electronic)
Oberwolfach Seminars
ISBN 978-3-031-35549-3 ISBN 978-3-031-35550-9 (eBook)
https://doi.org/10.1007/978-3-031-35550-9

Mathematics Subject Classification: 35K55, 35Q35, 35R35, 53A10, 53C44, 65M60

This book is published under the imprint Birkhäuser, www.birkhauser-science.com by the registered company Springer Nature Switzerland AG
The registered company address is: Gewerbestrasse 11, 6330 Cham, Switzerland

Paper in this product is recyclable.

Preface

These lecture notes are dedicated to the mathematical modeling, analysis, and computation of interfaces and free boundary problems appearing in geometry and in various applications, ranging from crystal growth, tumor growth, biological membranes to porous media, two-phase flows, fluid-structure interactions, and shape optimization. Classical methods from partial differential equations as well as from differential geometry, together with modern methods like the theory of maximal regularity or measure theoretic approaches, now allow for a systematic mathematical theory for interfaces and free boundary problems in many settings. Also, numerical methods based on parametric approaches, level sets, or phase fields are now mature enough to deal with interesting phenomena. However, in many applications, quite complex couplings between equations on the interface and equations in the surrounding bulk phases appear, which are still not well understood so far.

We first give an introduction to classical methods from differential geometry and systematically derive the governing equations from physical principles. Then we analyze parametric approaches to interface evolution problems and derive numerical methods which are thoroughly analyzed. In addition, implicit descriptions of interfaces such as phase field and level set methods are analyzed. Finally, we discuss numerical methods for complex interface evolutions and focus on two phase flow problems as an important example of such evolutions.

Some parts of the lecture notes have been first used by the first and the third authors in courses they gave to doctoral students of the DFG research training group 2339 *IntComSin: Interfaces, Complex Structures, and Singular Limits in Continuum Mechanics—Analysis and Numerics*. The complete material of this book has been presented at an Oberwolfach seminar in November 2022. We thank the Oberwolfach Research Institute for Mathematics (MFO) for giving us the opportunity to give the lecture series in the Oberwolfach seminar series. We thank all the participants of the seminar for actively taking part in the seminar. Due to their many suggestions for improvements, we were able to substantially enhance the presentation. We would like to thank Paul Hüttl, Jonas Haselböck, and Dennis Trautwein for proofreading. We would especially like to thank Eva Rütz and Vera Theus for typesetting parts of the notes and Jiří Minarčík for creating many of the figures in this book.

Erlangen, Germany — Eberhard Bänsch
Magdeburg, Germany — Klaus Deckelnick
Regensburg, Germany — Harald Garcke
Essen, Germany — Paola Pozzi

Contents

About the Authors

Eberhard Bänsch studied Mathematics at the University of Bonn and finished his PhD in 1990 as a student of Hans Wilhelm Alt. After that, he went to the University of Freiburg, where he received his habilitation in 1998. The same year, he went to Bremen for his first professorship. In 2000, he accepted an offer from Berlin, where he was appointed full professor at the Free University of Berlin and at the same time head of the research group Numerical Mathematics and Scientific Computing at the Weierstrass-Institute. He was also a member of the research center MATHEON in Berlin. Since 2004, he is full professor at the Friedrich-Alexander-University at Erlangen-Nuremberg.

The scientific interest of Eberhard Bänsch lies in the field of numerical analysis and scientific computing for free boundary and interface problems, in particular in computational fluid dynamics.

Copyright: Eberhard Bänsch.

Klaus Deckelnick studied mathematics at the University of Bonn, where he received his PhD in 1990. He subsequently worked as a scientific assistant at the University of Freiburg completing his habilitation in 1996. In 1998, he joined the University of Sussex where he had previously spent one year as a research fellow. Since 2002, he is a professor at the University of Magdeburg.

Deckelnick works on the analysis and numerical analysis of nonlinear partial differential equations with a particular focus on geometric evolution equations.

Courtesy: Archives of the Mathematisches Forschungsinstitut Oberwolfach.

Harald Garcke studied mathematics at the University of Bonn and finished his PhD in 1993 as a student of Hans Wilhelm Alt. With an ESF fellowship, he was in 1993/94 post-doc with Charles M. Elliott at the University of Sussex and from 1994 he was scientific assistant in Bonn where he finished his habilitation in 2000. In 2001, he got offers for professor positions at the Universities of Regensburg and Duisburg. Since 2002, he is full professor at the University of Regensburg where he was dean of the Mathematics Department from 2005 to 2007. He is DFG liaison officer at the University of Regensburg since 2011.

Harald Garcke works on nonlinear partial differential equations, free boundary problems, phase field equations, numerical analysis, and geometric evolution equations.

Courtesy: Archives of the Mathematisches Forschungsinstitut Oberwolfach.

Paola Pozzi received her undergraduate education in Mathematics at the University of Bologna in Italy. As a recipient of an Australian National University PhD Scholarship and an International Postgraduate Research Scholarship, she could then move to Canberra to work on her dissertation under the supervision of John Hutchinson at the Australian National University. After receiving her PhD in 2004 and a postdoctoral fellowship at the ANU Mathematical Science Institute, she moved to Germany to work as a scientific assistant in the research group of Gerhard Dziuk at the University of Freiburg. In 2009–2010, she was appointed for 6 months as a substitute professor (Vertretungsprofessur) at the Technische Universität München. In 2011, she got offers for professor positions at the Freie Universität in Berlin and at the University of Duisburg-Essen. She has been working as a professor at the University of Duisburg-Essen since 2011 and has acted as vice-dean (Prodekanin) of the Faculty of Mathematics between 2018 and 2020.

Pozzi works on analysis and numerical analysis for free boundary problems, geometric evolution equations, and nonlinear partial differential equations.

Copyright: UDE/Frank Preuß.

Introduction

1

Abstract

In the introduction we state some examples of applications in which interfaces appear and for which we will analyze mathematical approaches in more detail in later parts of the lecture notes. In particular, we will discuss grain boundary motion, melting and solidification, flow problems with interfaces and biomembranes.

This lecture series will give an introduction to mathematical methods for dealing with interfaces. Interfaces separating different physical states appear in various applications, ranging from crystal growth, tumor growth, biological membranes to porous media, two-phase flows, fluid-structure interactions, and shape optimization. In order to mathematically deal with interfaces, first of all, one has to decide how to describe the interface. Parametrizations, level set approaches, volume of fluid methods, measure theoretic approaches and phase field methods are classical ways to represent interfaces.

Using a parametric approach for interface evolution typically leads to fully nonlinear or at least quasi-linear partial differential equations whose exploration requires knowledge of the theory of abstract evolution equations and maximal regularity. Weak approaches for interfaces often involve measure theoretic methods such as varifold theory and approaches that use Caccioppoli sets. An advantage of these approaches is that they allow for topological changes which are not directly possible in sharp interface methods based on parametrizations. In phase field methods, the interface is described as a diffuse interfacial layer. The governing equations typically allow for quite smooth solutions, which is beneficial both for the analysis and for the numerics. Although phase field methods can be derived using classical thermodynamical principles, it is important to relate them to classical sharp interface descriptions, and many analytical questions in this context are still open. In addition, it is sometimes possible to describe interfaces with the help of obstacle

E. Bänsch et al., *Interfaces: Modeling, Analysis, Numerics*,
Oberwolfach Seminars 51, https://doi.org/10.1007/978-3-031-35550-9_1

problems or non-smooth complementarity conditions. Here, a suitable relaxation is often needed, and a limit analysis as well as the construction of robust approximation schemes are issues of research.

Different representations of the interface go with different numerical approaches. Recently, classical parametric descriptions of the interface have come back into the focus of the international research community because new ideas entered the field, yielding in particular, a better mesh quality. Phase field approaches and level set methods, in contrast, allow for an implicit way to treat topological changes.

In this introduction we will first state some examples of interfaces which we will analyze in more detail later. Chapter 2 will introduce basic facts about surfaces which will be needed in order to describe interfaces. In Chap. 3 we will derive mathematical models involving interfaces. Chapter 4 discusses analytical and numerical tools for dealing with parametric approaches for interfaces and in Chap. 5, methods for implicit approaches to interfaces will be introduced. Finally, Sect. 6 deals with numerical methods for complex interface evolutions and, in particular, the case of two phase flows will be considered.

1.1 Grain Boundary Motion

A grain boundary is an interface where crystals of different orientations meet. The crystals on each side of the interface are identical except in orientation. Solid materials consisting of crystals with different orientations are called polycrystalline materials. A situation in which polycrystal regions with different crystallographic orientations appear is sketched in Fig. 1.1. These different regions are separated by interfaces, which evolve in time by the evolution law

$$\text{normal velocity} = \text{mean curvature}$$

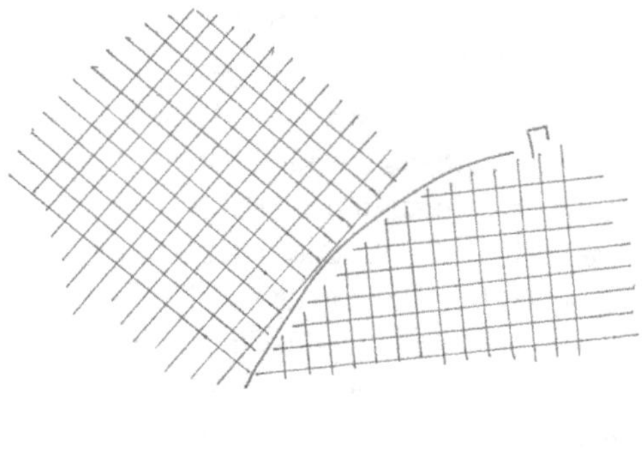

Fig. 1.1 A grain boundary in a polycrystal

or in a formula

$$V = \kappa\,.$$

The precise meaning of V and κ will be given in Chap. 2.

1.2 Melting and Solidification

In more complex mathematical models for interfaces not only the interface itself is an unknown but also quantities (functions) away from the interface (in the *bulk*). A typical example is the melting ice cube in your whiskey glass or a growing snow crystal, see Fig. 1.2. Here, functions describing the temperature or the concentration of water are unknowns, as well as the interface itself. The temperature, for example, has to solve a heat equation in an unknown domain. One hence speaks of a free boundary problem. The Stefan problem for melting and solidification is a typical problem in this context. We refer to Fig. 1.3 for a sketch of the mathematical setting.

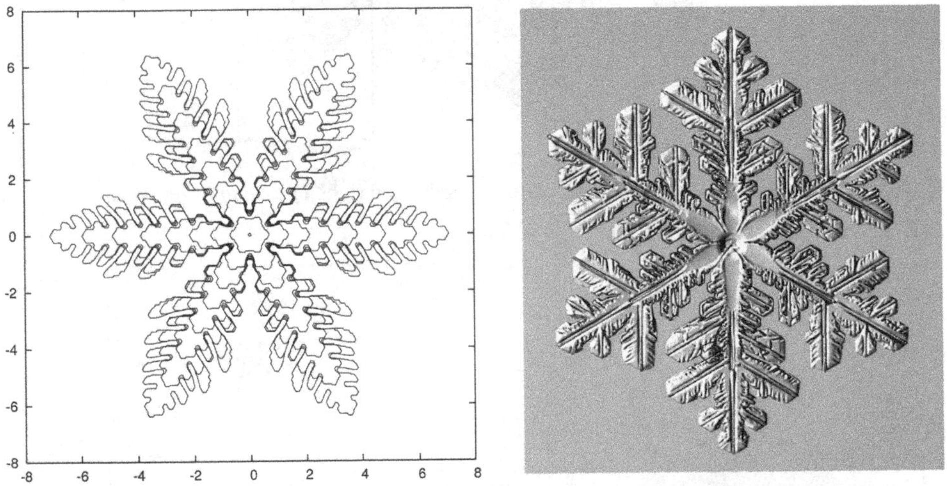

Fig. 1.2 Dendritic growth of a snow crystal. On the left a computer simulation of a snow crystal (by Robert Nürnberg, Trento, see [17]), and on the right a real snow crystal (photo courtesy of Kenneth Libbrecht, Caltech)

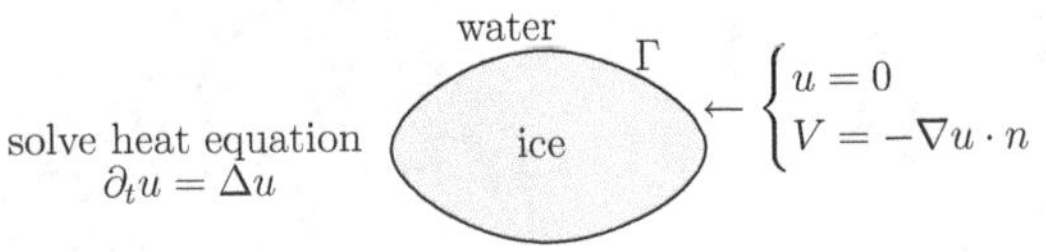

Fig. 1.3 We sketch a simple version of the Stefan problem. In more complete models the conditions at the interface involve the curvature of the interface

1.3 Flow Problems with Interfaces

Bubbles, drops, and particles are of fundamental importance in a multitude of physical, chemical, and biological processes and they also have important applications in industry. We mention rainfall, boiling, sprays, blood flow and lubrication. Compared to solidification the problems involving flows become more complex as the Navier–Stokes equation has to be solved on a domain with a moving boundary. Figures 1.4 and 1.5 show typical scenarios for fluid with an interface.

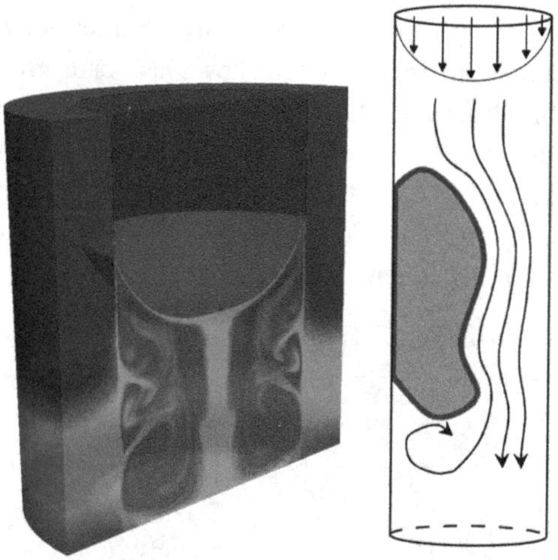

Fig. 1.4 Left a capillary driven flow under microgravity and right a growing thrombus in blood flow

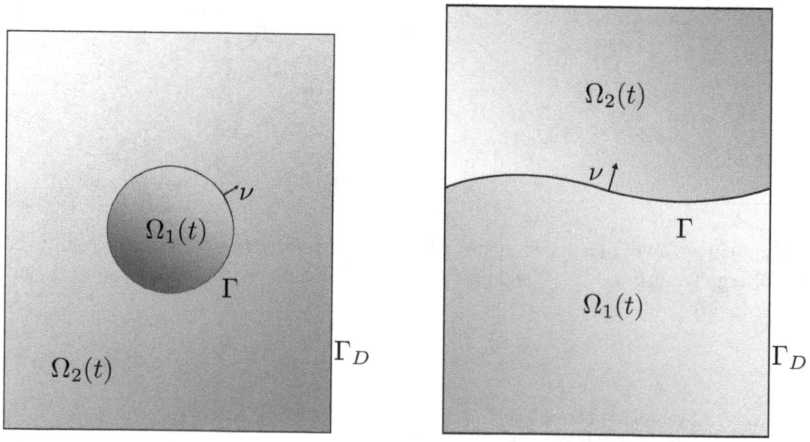

(a) (b)

Fig. 1.5 Examples for two-phase flow. (**a**) Rising drop. (**b**) Container with two fluids

1.4 Curvature Energies and Biomembranes

Curvature energies for surfaces in Euclidean space lead to fundamental geometric functionals which play an important role in differential geometry, in image processing, in surface restoration and in many physical models for beams, shells and membranes. They also appear in biology as, for example, the free energy of cell membranes contains curvature energies. This is important for example in the study of red blood cells, see, e.g., Fig. 1.6. The simplest form of a curvature energy is given by

$$\frac{1}{2}\int_M \kappa^2(x)\, d\mathcal{H}^m(x) \tag{1.1}$$

where κ is the mean curvature and M is the surface for which one wants to compute the energy.

Fig. 1.6 A red blood cell taken from https://commons.wikimedia.org/wiki/File:BloodCellState_080_image_of_a_state_of_a_blood_cell.png?uselang=de

Some Notions from Differential Geometry

2

Abstract

After recalling some basic notions from differential geometry, and reviewing the fundamental concepts of curvature, differentiation and integration on manifolds, we study surfaces evolving in time. In particular, we introduce the definitions of material derivative and normal time derivative, and derive useful versions of transport theorems that will be frequently used in subsequent chapters.

2.1 What Is a Surface?

Let us first recall the definition of a manifold M in $\mathbb{R}^n$ of dimension $m \leq n$ (see Fig. 2.1). In order to avoid too many technicalities, in what follows we will be rather informal. Mathematically precise formulations can be found for instance in [48, 105].

- 1. Definition: Representation by parametrizations
  A set $M \subset \mathbb{R}^n$ is a (smooth) m-dimensional manifold, if M can be locally represented as follows. For arbitrary $x \in M$ there exists an open neighborhood $V \subset \mathbb{R}^n$ of x, as well as an open set $U \subset \mathbb{R}^m$, and a mapping $F : U \to M$ such that

  $$F : U \to M \cap V \text{ is bijective,}$$

  F is smooth, F^{-1} is continuous and for all $\hat{x} \in U$ the differential $DF(\hat{x}) \in \mathbb{R}^{n \times m}$ has maximal rank. The map F is called a **local parametrization** (see Fig. 2.2).

E. Bänsch et al., *Interfaces: Modeling, Analysis, Numerics*,
Oberwolfach Seminars 51, https://doi.org/10.1007/978-3-031-35550-9_2

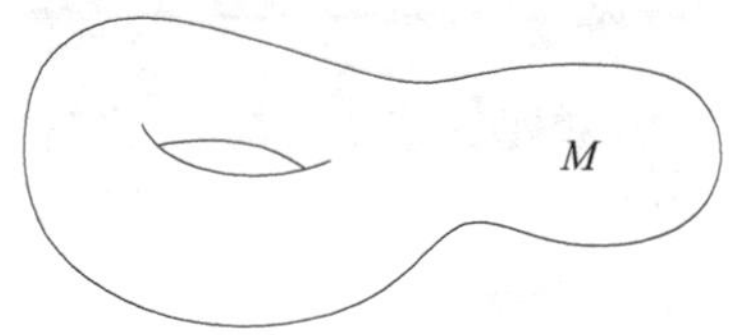

Fig. 2.1 A manifold

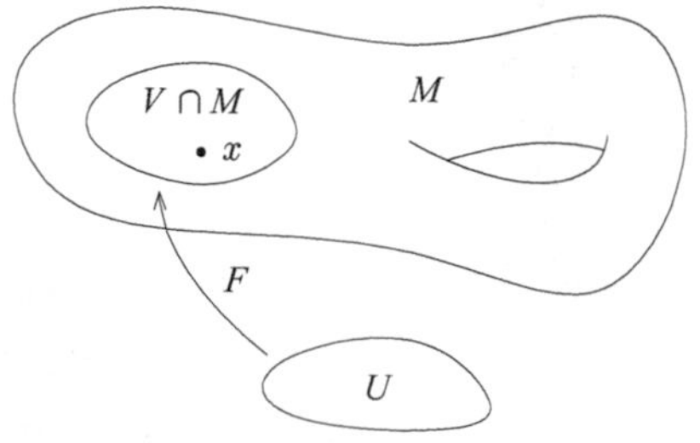

Fig. 2.2 Representation by a local parametrization

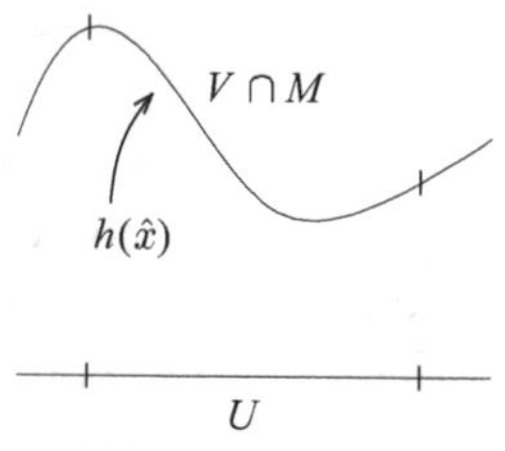

Fig. 2.3 Representation as a graph

- 2. Definition: Representation as a graph
  After a possible rotation of the coordinate system, M is locally a graph of a function h, i.e., for arbitrary $x \in M$ there exists a rotation of the coordinate system such that after rotation there exists an open neighborhood $V \subset \mathbb{R}^n$ of x, an open set $U \subset \mathbb{R}^m$ and a smooth function $h : U \to \mathbb{R}^{n-m}$ with

$$V \cap M = \{(\hat{x}, h_1(\hat{x}), ..., h_{n-m}(\hat{x})) \mid \hat{x} \in U\},$$

  (see Fig. 2.3).
- 3. Definition: Representation as a level-set
  For arbitrary $x \in M$ there exists an open neighborhood $V \subset \mathbb{R}^n$ of x and a smooth $\phi : V \to \mathbb{R}^{n-m}$ with $D\phi(y) \in \mathbb{R}^{(n-m)\times n}$ surjective for all $y \in V$ such that

$$V \cap M = \{y \in V \mid \phi(y) = 0\}.$$

Note that the three definitions are equivalent. Above we wrote smooth manifold. In fact we could more precisely speak of C^k-manifolds, with $k \in \mathbb{N}$, $k \geq 1$, if the involved functions are C^k.

2.2 Integration and Differentiation on a Surface

In what follows we denote by $M \subset \mathbb{R}^n$ an m-dimensional smooth manifold. According to Definition 1, M can be locally represented by $F_l : U_l \to \mathbb{R}^n$ with l indicating the different parametrizations. Let the local pieces $F_l(U_l) \subset M$ be mutually disjoint and such that their union cover M up to measure zero. Then the **integral** of f over M is defined as

$$\int_M f(x) d\mathcal{H}^m(x) := \sum_l \int_{U_l} (f \circ F_l)(\hat{x}) \sqrt{g_l(\hat{x})}\, d\hat{x}, \tag{2.1}$$

where $g_{l,ij}(\hat{x}) := \partial_i F_l(\hat{x}) \cdot \partial_j F_l(\hat{x})$ is the metric tensor and $g_l = \det(g_{l,ij})_{i,j}$. In the more common case that the $F_l(U_l)$ overlap, we use a partition of unity.

Next, given $x \in M$ let $F : U \to M$ be a local parametrization of M with $x = F(\hat{x})$. Then the m-dimensional subspace

$$T_M(x) := \text{span}\{\partial_{\hat{x}_j} F(\hat{x}) \mid j = 1, ..., m\}$$

of $\mathbb{R}^n$ is called the tangent space to M at x. Note that $T_M(x)$ is independent of the special choice of F (Fig. 2.4).

We say that a function $f : M \to \mathbb{R}$ is differentiable near $x \in M$ if there exists a local parametrization $F : U \to \mathbb{R}^n$ of M with $x = F(\hat{x})$ such that $f \circ F$ is differentiable near $\hat{x}$. In this case we define its **tangential gradient** by

$$\nabla_M f(x) := g^{ij}(\hat{x}) \partial_{\hat{x}_j} (f \circ F)(\hat{x}) \partial_{\hat{x}_i} F(\hat{x}) \in T_M(x) \tag{2.2}$$

with $T_M(x)$ the tangent space at x. Note that we have used the **Einstein convention** in the formula above: *repeated indices are summed up*. The somewhat complicated notation of the tangential gradient will become clearer later. Here,

$$g^{ij} = (g_{ij})^{-1}_{ij}, \qquad i, j = 1, ..., m$$

is the inverse of the metric tensor. It can be shown that if $f \circ F$ is differentiable near x for some parametrization F the same is true for any local parametrization of M near x and that $\nabla_M f(x)$ is independent of the special choice of F.

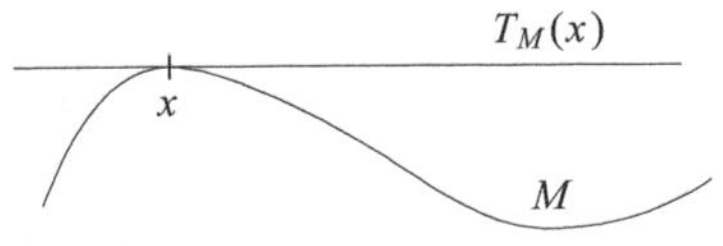

Fig. 2.4 Manifold M and tangent space $T_M(x)$

For a vector field $v : M \to \mathbb{R}^n$ which is differentiable near $x \in M$ we define its **surface Jacobian** $\nabla_M v(x) \in \mathbb{R}^{n\times n}$ by

$$(\nabla_M v(x))_{ij} := (\nabla_M v_i(x))_j, \qquad 1 \le i, j \le n \tag{2.3}$$

while its **surface divergence** is given by

$$\nabla_M \cdot v(x) := \mathrm{div}_M v(x) := \mathrm{Tr}(\nabla_M v(x)). \tag{2.4}$$

In the above we think of the tangential gradient $\nabla_M f$ as a *column* vector, whereas for a vector valued function v the i-th *row* of the surface Jacobian is the transposed gradient of the i-th component.

Note that

$$\forall w \in (T_M(x))^\perp : \quad (\nabla_M v(x) w)_i = (\nabla_M v(x))_{ij} w_j = \nabla_M v_i(x) \cdot w = 0 \tag{2.5}$$

since $\nabla_M v_i(x) \in T_M(x)$. In particular we see that $\mathrm{rank}\nabla_M v(x) \le m$.

The next result shows how to compute the tangential gradient in the case that f has an extension to an open neighbourhood of x.

Proposition 2.2.1 *Let $f : M \to \mathbb{R}$ be differentiable near $x \in M$ and $\tilde{f} : V \to \mathbb{R}$ differentiable in an open neighbourhood of x with $\tilde{f}(y) = f(y)$ for all $y \in V \cap M$. Then we have*

$$\nabla_M f(x) = P(x)\nabla\tilde{f}(x)$$

where $P(x)$ denotes the orthogonal projection of $\mathbb{R}^n$ onto $T_M(x)$.

Likewise, let $v : M \to \mathbb{R}^n$ be differentiable near x and $\tilde{v} : V \to \mathbb{R}^n$ differentiable with $\tilde{v}(y) = v(y)$ for all $y \in V \cap M$. Then we have

$$\nabla_M v(x) = \nabla\tilde{v}(x)P(x),$$

i.e., $(\nabla_M v(x))_{ij} = \partial_k \tilde{v}_i(x) P_{kj}(x)$, $1 \le i, j \le n$.

Proof To begin, note that $P(x) = I - \sum_{l=1}^{n-m} \nu_l(x) \otimes \nu_l(x)$, where $\{\nu_1(x), \ldots, \nu_{n-m}(x)\}$ is an orthonormal basis of the orthogonal complement of $T_M(x)$.

Let $F : U \to M$ be a local parametrization of M with $x = F(\hat{x})$. For $k = 1, \ldots, m$ we have

$$\begin{aligned}\nabla_M f(x) \cdot \partial_{\hat{x}_k} F(\hat{x}) &= g^{ij}(\hat{x}) \partial_{\hat{x}_j}(f \circ F)(\hat{x}) \partial_{\hat{x}_i} F(\hat{x}) \cdot \partial_{\hat{x}_k} F(\hat{x}) = g^{ij}(\hat{x}) g_{ik}(\hat{x}) \partial_{\hat{x}_j}(\tilde{f} \circ F)(\hat{x}) \\ &= \partial_{\hat{x}_k}(\tilde{f} \circ F)(\hat{x}) = \nabla \tilde{f}(x) \cdot \partial_{\hat{x}_k} F(\hat{x}) = (P(x) \nabla \tilde{f}(x)) \cdot \partial_{\hat{x}_k} F(\hat{x}),\end{aligned}$$

where we have used the chain rule and the fact that $\partial_{\hat{x}_k} F(\hat{x}) \in T_M(x)$. Since

$$\nabla_M f(x) \cdot w = 0 = P(x) \nabla \tilde{f}(x) \cdot w \qquad \text{for all } w \in (T_M(x))^{\perp},$$

we deduce that $\nabla_M f(x) = P(x) \nabla \tilde{f}(x)$. Applying this result to the components of v, recalling the definition of the surface Jacobian and using the symmetry of $P(x)$ we obtain the remaining formula. □

There is a second order operator, the Laplace–Beltrami operator, that plays a similarly fundamental role on surfaces as the Laplace operator in $\mathbb{R}^n$ and is its generalization to manifolds.

Definition 2.2.2 (Laplace–Beltrami Operator) For $f : M \to \mathbb{R}$ smooth we define

$$\Delta_M f := \nabla_M \cdot (\nabla_M f) \tag{2.6}$$

which is called the **Laplace–Beltrami operator**.

Remark

(a) Let $F : U \to M$ be a local parametrization of M, $g_{ij} = \partial_{\hat{x}_i} F \cdot \partial_{\hat{x}_j} F$, $i, j = 1, ..., m$, and $g = \det(g_{ij})$, then Δ_M can be written as (see Exercise 7.1)

$$\Delta_M f(x) = \frac{1}{\sqrt{g}} \partial_{\hat{x}_i}(\sqrt{g}\, g^{ij} \partial_{\hat{x}_j}(f \circ F))(\hat{x}), \quad \text{with} \quad x = F(\hat{x}). \tag{2.7}$$

(b) Let $M \subset \mathbb{R}^n$ be a one-dimensional manifold, in other words a curve, and $F : I \to M$ ($I \subset \mathbb{R}$ an interval) a parametrization of M. Then we have for a smooth function $f : M \to \mathbb{R}$

$$\Delta_M f(x) = \frac{1}{|F_{\hat{x}}(\hat{x})|}\Big(\frac{(f \circ F)_{\hat{x}}(\hat{x})}{|F_{\hat{x}}(\hat{x})|}\Big)_{\hat{x}}, \quad x = F(\hat{x}).$$

In the special case that F is a parametrization by arc length, i.e. $|F_{\hat{x}}(\hat{x})| = 1,\ \hat{x} \in I$, the above expression simplifies to

$$\Delta_M f(x) = (f \circ F)_{\hat{x}\hat{x}}(\hat{x}), \quad \hat{x} \in I.$$

2.3 Weingarten Map

In what follows, we are always considering *hypersurfaces*, i.e. manifolds of dimension $m = n - 1$ (one may also say M has co-dimension 1), if not otherwise stated. In what follows, we denote hypersurfaces by Γ. Then for each $x \in \Gamma$ the tangent space $T_\Gamma(x)$ is an $(n-1)$-dimensional subspace of $\mathbb{R}^n$ and there exists an open neighbourhood V of x and a smooth unit normal field $\nu : V \cap \Gamma \to \mathbb{R}^n$ to Γ, i.e.,

$$\nu(y) \perp T_\Gamma(y), \quad |\nu(y)| = 1 \quad \text{for all } y \in V \cap \Gamma.$$

In order to see this recall that $V \cap \Gamma = \{y \in V \mid \phi(y) = 0\}$ according to Definition 3 in Sect. 2.1 with a smooth function $\phi : V \to \mathbb{R}$ satisfying $\nabla\phi(y) \neq 0,\ y \in V$. A possible choice then is $\nu = \tilde{\nu}_{|\Gamma}$, where

$$\tilde{\nu}(y) = \frac{\nabla\phi(y)}{|\nabla\phi(y)|}. \tag{2.8}$$

Definition 2.3.1 (Weingarten Map and 2nd Fundamental Form) For $x \in \Gamma$ the map

$$H(x) := -\nabla_\Gamma \nu(x) \in \mathbb{R}^{n\times n} \text{ is called the Weingarten map at } x.$$

The associated quadratic form

$$(v, u) \mapsto v \cdot (H(x)u)$$

is called the second fundamental form.

Proposition 2.3.2 *The Weingarten map $H(x)$ is symmetric and $H(x)\nu(x) = 0$, i.e. $\nu(x) \in ker(H(x))$.*

Proof In view of Proposition 2.2.1 and (2.8) we have

$$H(x)_{ij} = -(\nabla_\Gamma \nu(x))_{ij} = -\partial_k \tilde{\nu}_i(x) P_{kj}(x),$$

where $\tilde{\nu}_i(y) = \frac{\partial_i \phi(y)}{|\nabla \phi(y)|}$ and $P(x) = I - \nu(x) \otimes \nu(x)$ is the orthogonal projection onto $T_\Gamma(x)$. Clearly,

$$\partial_k \tilde{\nu}_i(x) = \frac{\partial_k \partial_i \phi(x)}{|\nabla \phi(x)|} - \frac{\partial_i \phi(x)\, \partial_l \phi(x)\, \partial_k \partial_l \phi(x)}{|\nabla \phi(x)|^3},$$

so that (omitting the x for a moment)

$$\begin{aligned} H_{ij} &= -(\delta_{jk} - \frac{\partial_j \phi}{|\nabla \phi|} \frac{\partial_k \phi}{|\nabla \phi|})(\frac{\partial_k \partial_i \phi}{|\nabla \phi|} - \frac{\partial_i \phi\, \partial_l \phi\, \partial_k \partial_l \phi}{|\nabla \phi|^3}) \\ &= -\frac{\partial_j \partial_i \phi}{|\nabla \phi|} + \frac{\partial_i \phi\, \partial_l \phi\, \partial_j \partial_l \phi}{|\nabla \phi|^3} + \frac{\partial_j \phi\, \partial_k \phi\, \partial_k \partial_i \phi}{|\nabla \phi|^3} - \frac{\partial_i \phi\, \partial_j \phi\, \partial_k \phi\, \partial_l \phi\, \partial_k \partial_l \phi}{|\nabla \phi|^5} \end{aligned}$$

which together with Schwarz's theorem implies that $H(x)_{ij} = H(x)_{ji}$ for all $1 \le i, j \le n$. Finally, we deduce from (2.5) that $H(x)\nu(x) = -(\nabla_\Gamma \nu(x))\nu(x) = 0$, since $\nu(x) \in (T_\Gamma(x))^\perp$. □

Since $H(x)$ is symmetric, there is an orthonormal basis of eigenvectors with real eigenvalues $\kappa_1(x), ..., \kappa_n(x)$. Note that at least one of the eigenvalues is 0, since $\nu(x) \in ker(H(x))$. We denote this eigenvalue by $\kappa_n(x)$, i.e., $\kappa_n(x) = 0$. The quantity

$$\kappa(x) := \sum_{i=1}^{n-1} \kappa_i(x) = \sum_{i=1}^{n} \kappa_i(x) = \mathrm{Tr}(H(x)) = -\nabla_\Gamma \cdot \nu(x) \tag{2.9}$$

is called *mean curvature* of Γ at x (strictly speaking $\frac{1}{n-1} \sum_{i=1}^{n-1} \kappa_i(x)$ is the mean curvature; however, dropping the factor $1/(n-1)$ is more convenient). As the trace is invariant under basis changes, we obtain that $\kappa(x)$ is independent of the coordinate system.

The Weingarten map H and in turn the mean curvature κ depend on the orientation of ν. A change of the orientation $\nu \mapsto -\nu$ leads to $H \mapsto -H$ and $\kappa \mapsto -\kappa$.

Proposition 2.3.3 *Let $x \in \Gamma$ and suppose that there exists an open neighborhood V of x such that $V \cap \Gamma = \{y \in V \mid \phi(y) = 0\}$. Let the unit normal field ν on $V \cap \Gamma$ be given by $\nu = \tilde{\nu}_{|V\cap\Gamma}$, with $\tilde{\nu} : V \to \mathbb{R}^n$ defined as $\tilde{\nu}(y) = \dfrac{\nabla\phi(y)}{|\nabla\phi(y)|}$. Then $\kappa(x) = -\nabla \cdot \dfrac{\nabla\phi(x)}{|\nabla\phi(x)|}$.*

Proof This is left as an exercise. □

We have the following useful identities.

Proposition 2.3.4 *Let $\chi : \mathbb{R}^n \to \mathbb{R}^n$ be the identity map, i.e., $\chi(x) = x$ for all $x \in \mathbb{R}^n$. Then we have for $x \in \Gamma$:*

(i) $\nabla_\Gamma \chi(x) = P(x) = I - \nu(x) \otimes \nu(x)$;
(ii) $\Delta_\Gamma \chi(x) = \kappa(x)\nu(x)$.

The vector $\kappa(x)\,\nu(x)$ is called **curvature vector** *at $x \in \Gamma$.*

Proof

(i) We compute with the help of Proposition 2.2.1

$$(\nabla_\Gamma \chi(x))_{ij} = \partial_k \chi_i(x) P_{kj}(x) = \delta_{ik} P_{kj}(x) = P_{ij}(x).$$

(ii) Using (i) we deduce for $i = 1, \dots, n$

$$\begin{aligned}\Delta_\Gamma \chi_i(x) &= \nabla_\Gamma \cdot \nabla_\Gamma \chi_i(x) = [\nabla_\Gamma(\nabla_\Gamma \chi_i)_j]_j(x) = [\nabla_\Gamma(\delta_{ij} - \nu_i\nu_j)]_j(x)\\ &= -(\nabla_\Gamma \nu_i(x))_j \nu_j(x) - (\nabla_\Gamma \nu_j)(x))_j \nu_i(x) = -\nabla_\Gamma \cdot \nu(x)\,\nu_i(x) = \kappa(x)\nu_i(x),\end{aligned}$$

so that $\Delta_\Gamma \chi(x) = \kappa(x)\nu(x)$. □

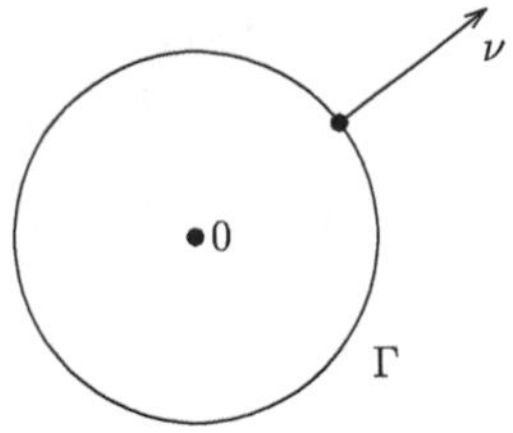

Fig. 2.5 $\Gamma = \partial B_R(0)$

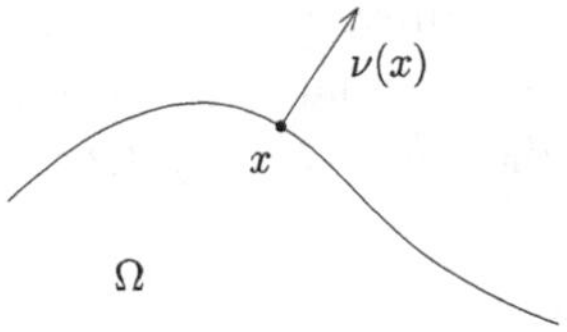

Fig. 2.6 Domain Ω and outer normal ν

Example For $R > 0$ we denote by $\Gamma := \mathbb{S}_R^{n-1} = \partial B_R(0) \subset \mathbb{R}^n$ the sphere with radius R (Fig. 2.5). For the unit normal on Γ pointing outwards we have $\nu = \tilde{\nu}_{|\Gamma}$, where $\tilde{\nu}(y) = \frac{y}{R}$. Thus we obtain with the help of Proposition 2.2.1 that

$$H(x) = -\nabla\tilde{\nu}(x)P(x) = -\frac{1}{R}P(x) = -\frac{1}{R}(I - \frac{x}{R} \otimes \frac{x}{R}), \quad x \in \Gamma.$$

Thus each vector $v \in T_\Gamma(x) \setminus \{0\}$ is an eigenvector for the eigenvalue $-\frac{1}{R}$, while $\frac{x}{R}$ is an eigenvector for $\kappa_n(x) = 0$. In particular we have that $\kappa(x) = -\frac{n-1}{R}$, $x \in \Gamma$.

2.4 Signed Distance Function and Canonical Extension

In this section we follow the presentation in [36]. Let $\Gamma \subset \mathbb{R}^n$ be a smooth, connected, compact and orientable (i.e. there exists a smooth unit normal field $\nu : \Gamma \to \mathbb{R}^n$) hypersurface. It follows from the Jordan–Brouwer separation theorem that there exists a bounded open set $\Omega \subset \mathbb{R}^n$ lying on one side of Γ and having Γ as its boundary (Fig. 2.6). Replacing ν with $-\nu$ if necessary we may assume that

$$\begin{cases} |\nu(x)| = 1, \\ \nu(x) \perp T_\Gamma(x), \\ x - \delta\nu(x) \in \Omega, \text{ if } \delta > 0 \text{ is sufficiently small.} \end{cases}$$

It can be shown, see Boyer and Fabrie [29] in Section 3.2 in Chapter III, that for small $\delta > 0$ in the "tubular" neighborhood

$$D_\delta := \{x \in \mathbb{R}^n \mid x = z + \eta\nu(z), \; z \in \Gamma, \; |\eta| < \delta\}$$

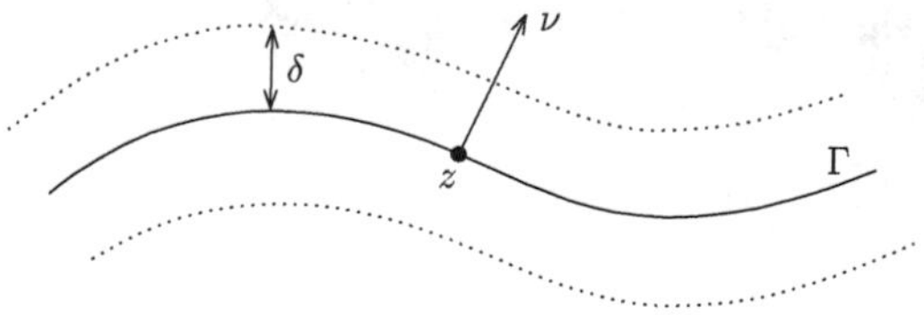

Fig. 2.7 "Tubular" neighborhood D_δ of Γ

of Γ there is a one-to-one relation between x and (z, η), i.e. for each $x \in D_\delta$ there exists one and only one $z \in \Gamma$ and $\eta \in (-\delta, \delta)$ with $x = z + \eta\nu(z)$ (Fig. 2.7). This means that there is a uniquely determined projection

$$\pi : D_\delta \to \Gamma,$$

$$x \mapsto z, \quad x = z + \eta\nu(z).$$

Moreover, π can be characterized by $\pi(x) = \operatorname{argmin}_{y\in\Gamma}|y - x|$.

Let $d_\Gamma : D_\delta \to \mathbb{R}$ be the *signed distance function*, that is (Fig. 2.8)

$$|d_\Gamma(x)| = \min_{y\in\Gamma} |x - y| \quad \text{for } x \in D_\delta$$

and

$$d_\Gamma(x) > 0 \text{ for } x = z + \eta\nu(z),\ \eta > 0, z \in \Gamma,$$

$$d_\Gamma(x) < 0 \text{ for } x = z + \eta\nu(z),\ \eta < 0, z \in \Gamma.$$

Proposition 2.4.1 *The functions π and d_Γ are smooth in D_δ and we have*

$$x = \pi(x) + d_\Gamma(x)\nu(\pi(x)), \quad x \in D_\delta, \tag{2.10}$$

$$\nabla d_\Gamma(x) = \nu(\pi(x)), \quad x \in D_\delta, \tag{2.11}$$

$$-\Delta d_\Gamma(x) = \kappa(x), \quad x \in \Gamma. \tag{2.12}$$

Proof We refer to Gilbarg, Trudinger [83], Section 14.6., and Boyer, Fabrie [29], Section 3.2 in Chapter III, for the first two identities. Applying Proposition 2.3.3 with $\phi = d_\Gamma$ and using the fact that $|\nabla d_\Gamma(y)| = |\nu(\pi(y))| = 1$ in D_δ we obtain

$$\kappa(x) = -\nabla \cdot \frac{\nabla d_\Gamma(x)}{|\nabla d_\Gamma(x)|} = -\nabla \cdot \nabla d_\Gamma(x) = -\Delta d_\Gamma(x), \quad x \in \Gamma.$$

□

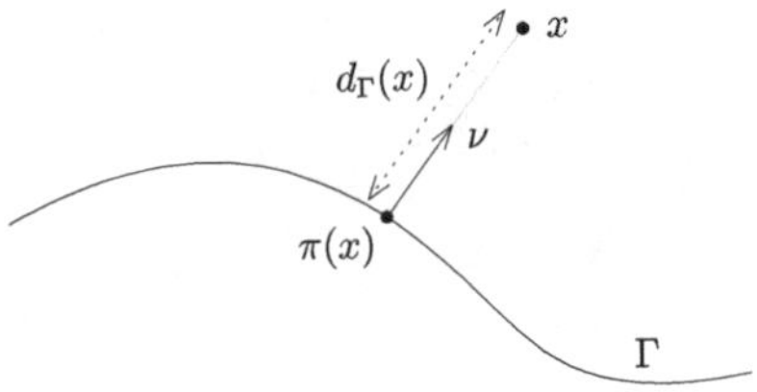

Fig. 2.8 Signed distance function d_Γ

With the help of the projection π, functions on Γ can be extended constantly in direction of ν, this means: for $f : \Gamma \to \mathbb{R}$ one can define

$$\hat{f}(x) := f(\pi(x)) \text{ for } x \in D_\delta. \tag{2.13}$$

We refer to $\hat{f}$ as the canonical extension of f. Note that for $x \in \Gamma$ we have that $\pi(x + s\nu(x)) = x$, so that $s \mapsto \hat{f}(x + s\nu(x))$ is constant. Therefore, $\nabla \hat{f}(x) \cdot \nu(x) = 0$ and we deduce from Proposition 2.2.1 that

$$\nabla_\Gamma f(x) = P(x)\nabla \hat{f}(x) = \nabla \hat{f}(x) \quad \text{for all} \quad x \in \Gamma. \tag{2.14}$$

In the same way vector valued functions $v : \Gamma \to \mathbb{R}^n$ are extended via

$$\hat{v}(x) := v(\pi(x)).$$

Any $v : \Gamma \to \mathbb{R}^n$ and any $\hat{v} : D_\delta \to \mathbb{R}^n$ can be decomposed in its tangential and normal part:

$$\hat{v}(x) = V(x)\hat{\nu}(x) + \hat{v}_\tau(x)$$

through

$$V(x) = \hat{v}(x) \cdot \hat{\nu}(x), \quad \hat{\nu}(x) = \nu(\pi(x)), \quad \hat{v}_\tau(x) = \hat{P}(x)\hat{v}(x),$$
$$\hat{P}(x) = P(\pi(x)) = I - \nu(\pi(x)) \otimes \nu(\pi(x)).$$

2.5 Integration by Parts on Manifolds

In this section we assume again that $\Gamma \subset \mathbb{R}^n$ is a smooth, connected, compact and orientable hypersurface. Let $M \subset \Gamma$ be **relatively open** in Γ, this means for all $x_0 \in M$ there exists an open set $U \subset \mathbb{R}^n$ with $x_0 \in U$ and $U \cap \Gamma \subset M$ (Fig. 2.9).

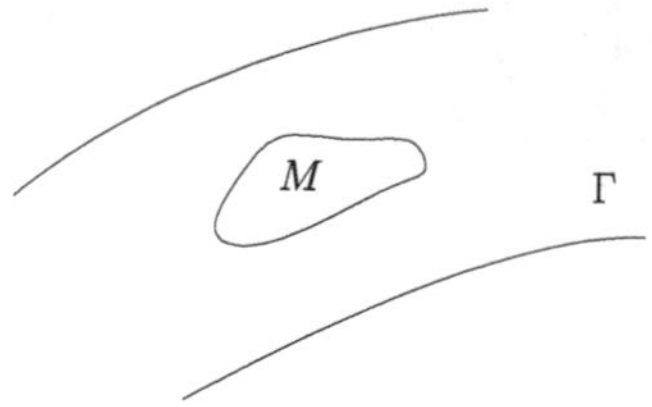

Fig. 2.9 M is relatively open in Γ

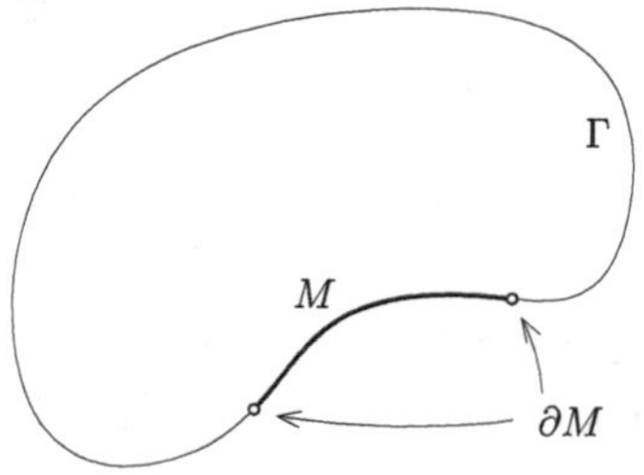

Fig. 2.10 Boundary ∂M

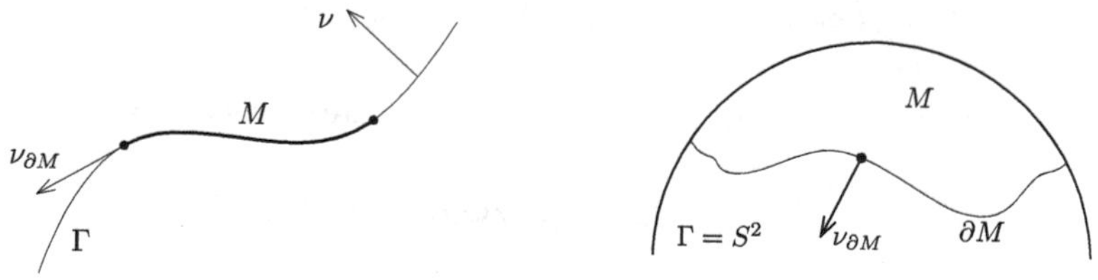

Fig. 2.11 Intrinsic normal $\nu_{\partial M}$

The **boundary** of M is defined as $\partial M := \{x \in \Gamma\backslash M \mid \exists x_k \in M,\ x_k \stackrel{k\to\infty}{\longrightarrow} x\}$. Let $z \in \partial M$. We say that a vector $v \in T_\Gamma(z)$ points into $\Gamma \setminus M$, if for every curve $\gamma : (-\epsilon, \epsilon) \to \mathbb{R}^n$ with $\gamma(\theta) \in \Gamma$, $\gamma(0) = z$ and $\gamma'(0) = v$ there exists $0 < \epsilon_1 \le \epsilon$ such that $\gamma(\theta) \in \Gamma \setminus M$ for $0 \le \theta \le \epsilon_1$ (Fig. 2.10).

The set M is called smooth in Γ, if $\partial M = \emptyset$ or ∂M is a smooth $(n-2)$-dimensional manifold and M lies on one side of ∂M in the sense that there exists a field $\nu_{\partial M} : \partial M \to \mathbb{R}^n$ with the following properties:

- $|\nu_{\partial M}(z)| = 1, \quad z \in \partial M,$
- $\nu_{\partial M}(z) \in T_\Gamma(z), \quad z \in \partial M,$
- $\nu_{\partial M}(z) \perp T_{\partial M}(z), \quad z \in \partial M$

and $\nu_{\partial M}(z)$ points into $\Gamma\setminus M$. The above conditions define a uniquely determined **intrinsic normal** field, which we call the outer unit conormal (Fig. 2.11).

Next we can generalize integration by parts from $\mathbb{R}^n$ to hypersurfaces, an important tool for the analysis on surfaces.

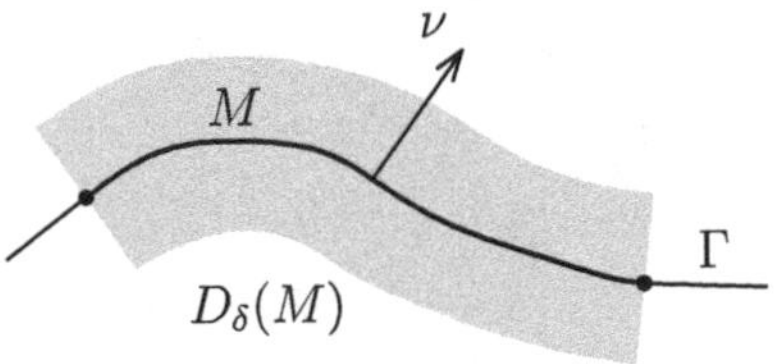

Fig. 2.12 The δ-neighborhood of M

Proposition 2.5.1 (Integration by Parts) *Let $M \subset \Gamma$ and $f : \Gamma \to \mathbb{R}$ be smooth. Then it holds:*

$$\int_M \nabla_\Gamma f \, d\mathcal{H}^{n-1} = -\int_M f\kappa\nu \, d\mathcal{H}^{n-1} + \int_{\partial M} f\nu_{\partial M} \, d\mathcal{H}^{n-2}.$$

Let $q : \Gamma \to \mathbb{R}^n$ be smooth. Then it holds:

$$\int_M \nabla_\Gamma \cdot q \, d\mathcal{H}^{n-1} = -\int_M \kappa\nu \cdot q \, d\mathcal{H}^{n-1} + \int_{\partial M} \nu_{\partial M} \cdot q \, d\mathcal{H}^{n-2}.$$

Sketch of the Proof *(See [36]; for a more classical proof see the proof of Theorem 21 in [23])* . Let us denote be d_Γ the signed distance function to Γ (cf. Sect. 2.4) and define

$$D_\delta(M) = \{x \in \mathbb{R}^n \mid x = z + \eta\nu(z), \; z \in M, \; |\eta| < \delta\} \subset \{-\delta < d_\Gamma < \delta\}$$

as the δ-neighborhood of M in $\mathbb{R}^n$, see Fig. 2.12. There exists $\delta_0 > 0$ such that d_Γ is smooth in $D_{\delta_0}(M)$. The idea is to approximate the integral on the manifold by integrals over the tubular neighborhood $D_\delta(M)$, perform integration by parts in the usual sense, divide by δ and then let δ tend to zero.

Let $g : D_{\delta_0}(M) \to \mathbb{R}$ be continuous. Then the coarea formula together with the fact that $|\nabla d_\Gamma(x)| = 1, x \in D_{\delta_0}(M)$ implies

$$\int_{D_\delta(M)} g \, dx = \int_{D_\delta(M)} g \, |\nabla d_\Gamma| \, dx = \int_{-\delta}^{\delta} \int_{D_\delta(M) \cap \{d_\Gamma = \eta\}} g \, d\mathcal{H}^{n-1} d\eta, \quad 0 < \delta \leq \delta_0,$$

so that

$$\int_M g \, d\mathcal{H}^{n-1} = \lim_{\delta \to 0} \frac{1}{2\delta} \int_{D_\delta(M)} g \, dx. \tag{2.15}$$

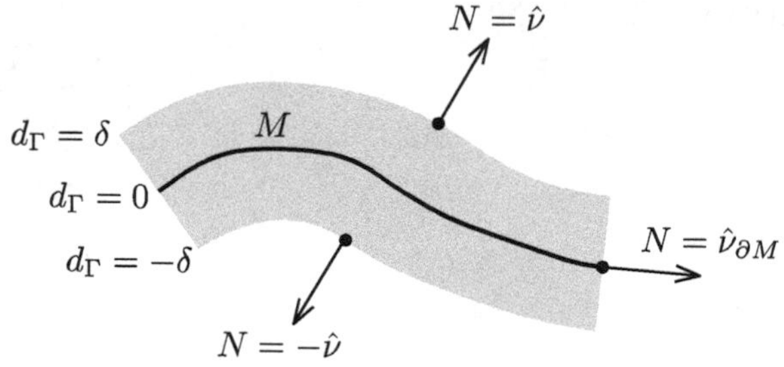

Let $\hat{f} : D_\delta(M) \to \mathbb{R}$ be the canonical extension of f and $\hat{\nu}(z) := \nabla d_\Gamma(z)$. Then we infer from (2.11) and (2.14) that

$$\hat{\nu}(x) = \nu(x), \quad \hat{P}(x)\nabla \hat{f}(x) = \nabla_\Gamma f(x), \quad x \in \Gamma,$$

where $\hat{P}(z) = I - \hat{\nu}(z) \otimes \hat{\nu}(z)$. Integration by parts together with (2.15) yields

$$\begin{aligned}
\Big(\int_M \nabla_\Gamma f \, d\mathcal{H}^{n-1}\Big)_i &= \lim_{\delta \to 0} \frac{1}{2\delta} \int_{D_\delta(M)} (\delta_{ij} - \hat{\nu}_i \hat{\nu}_j) \partial_j \hat{f} \, dx \\
&= \lim_{\delta \to 0} \frac{1}{2\delta} \Big[\int_{\partial D_\delta(M)} \hat{f}(\delta_{ij} - \hat{\nu}_i \hat{\nu}_j) N_j \, d\mathcal{H}^{n-1} \\
&\qquad - \int_{D_\delta(M)} \partial_j(\delta_{ij} - \hat{\nu}_i \hat{\nu}_j) \hat{f} \, dx \Big] = (*).
\end{aligned}$$

Here, N is the outer unit normal to $\partial D_\delta(M)$. In order to treat the first term we write

$$\partial D_\delta(M) = \{x = z \pm \delta\nu(z) \,|\, z \in M\} \cup \{x = z + \eta\nu(z) \,|\, z \in \partial M, |\eta| < \delta\} =: S_1^\delta \cup S_2^\delta.$$

For $x \in S_1^\delta$ one has: $N(x) = \pm\nu(\pi(x)) = \pm\hat{\nu}(x)$, which implies $(\hat{P}N)(x) = 0$. On the other hand, if $x = z + \eta\nu(z) \in S_2^\delta$, then $N(x) = \nu_{\partial M}(\pi(x))$ and hence $\hat{f}(x)\hat{P}(x)N(x) = f(\pi(x))\nu_{\partial M}(\pi(x))$, $x \in S_2^\delta$. For the second term in $(*)$ one computes $\partial_j \delta_{ij} = 0$ as well as

$$\partial_j(\hat{\nu}_i \hat{\nu}_j) = \partial_j(\partial_i d_\Gamma \partial_j d_\Gamma) = \partial_j \partial_i d_\Gamma \partial_j d_\Gamma + \Delta d_\Gamma \partial_i d_\Gamma = \Delta d_\Gamma \partial_i d_\Gamma,$$

since $\partial_j \partial_i d_\Gamma \partial_j d_\Gamma = \frac{1}{2}\partial_i |\nabla d_\Gamma|^2 = 0$. Thus (∗) together with (2.12) yields:

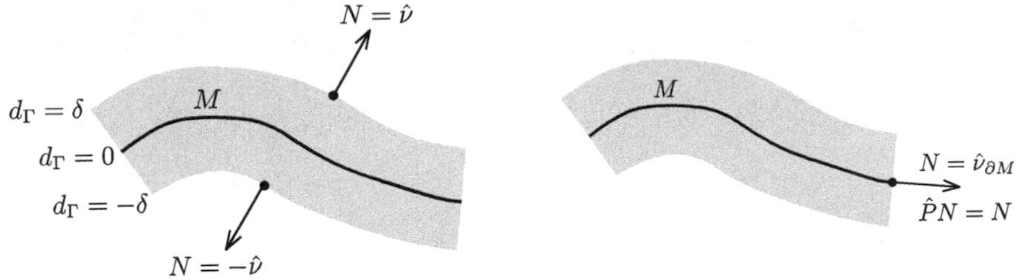

$$\Big(\int_M \nabla_\Gamma f \, d\mathcal{H}^{n-1}\Big)_i = \lim_{\delta \to 0} \frac{1}{2\delta}\Big[\int_{S_2^\delta} \hat{f}(\hat{P}N)_i \, d\mathcal{H}^{n-1} - \int_{D_\delta(M)} \partial_j(\delta_{ij} - \hat{\nu}_i \hat{\nu}_j)\hat{f} \, dx\Big]$$

$$= \int_{\partial M} f(\nu_{\partial M})_i \, d\mathcal{H}^{n-2} + \int_M f \Delta d_\Gamma \nu_i \, d\mathcal{H}^{n-1}$$

$$= \int_{\partial M} f(\nu_{\partial M})_i \, d\mathcal{H}^{n-2} - \int_M f \kappa \nu_i \, d\mathcal{H}^{n-1}$$

which proves the first identity of the proposition. The second identity follows from the first one by using its i'th component for q_i and then taking the sum over i. □

In the case that $M = \Gamma$, the boundary integral vanishes and we obtain the following useful identities:

Corollary 2.5.2

(a) Let $q : \Gamma \to \mathbb{R}^n$ be a ***tangential vector field*** *, i.e. $q(z) \in T_\Gamma(z)$ for all $z \in \Gamma$. Then it holds:*

$$\int_\Gamma \nabla_\Gamma \cdot q \, d\mathcal{H}^{n-1} = 0.$$

(b) Let χ be the identity map and $f : \Gamma \to \mathbb{R}$ be smooth. Then we have for $i = 1, \ldots, n$:

$$\int_\Gamma \nabla_\Gamma \chi_i \cdot \nabla_\Gamma f \, d\mathcal{H}^{n-1} = -\int_\Gamma \kappa f \nu_i \, d\mathcal{H}^{n-1}. \tag{2.16}$$

(continued)

Likewise there holds for a smooth vector field $v : \Gamma \to \mathbb{R}^n$

$$\int_\Gamma \nabla_\Gamma \chi : \nabla_\Gamma v \, d\mathcal{H}^{n-1} = -\int_\Gamma \kappa v \cdot \nu \, d\mathcal{H}^{n-1}. \tag{2.17}$$

Proof

(a) This follows from the second relation in Proposition 2.5.1, since $\partial\Gamma = \emptyset$ and $q(z) \cdot \nu(z) = 0$, $z \in \Gamma$.

(b) For the tangential vector field $q = f \nabla_\Gamma \chi_i$ we have that

$$\nabla_\Gamma \cdot q = \nabla_\Gamma f \cdot \nabla_\Gamma \chi_i + f \Delta_\Gamma \chi_i = \nabla_\Gamma f \cdot \nabla_\Gamma \chi_i + f \kappa \nu_i$$

in view of Proposition 2.3.4(ii). Hence (2.16) follows immediately from (a), while (2.17) is obtained by applying (2.16) to the components of v. □

2.6 Evolving Surfaces

We are mainly interested in interfaces which evolve in time. We consider a physical process happening in a domain $\Omega \subset \mathbb{R}^n$ in a time interval $(0, T)$. Assume two phases appear which are separated by an interface Γ. Now we define some geometric quantities related to the evolving interface Γ.

Definition 2.6.1 A family of sets $(\Gamma(t))_{t\in(0,T)}$ is called a smoothly evolving family of hypersurfaces in $\mathbb{R}^n$, if

(i) $\Gamma := \{(t, x) \in \mathbb{R} \times \mathbb{R}^n \,|\, t \in (0, T), x \in \Gamma(t)\}$ is a smooth hypersurface in $\mathbb{R} \times \mathbb{R}^n$,

(ii) the tangent spaces $T_\Gamma(t, x)$ of Γ are never space-like, i.e.,

$$T_\Gamma(t, x) \neq \{0\} \times \mathbb{R}^n \quad \text{for all } (t, x) \in \Gamma\,.$$

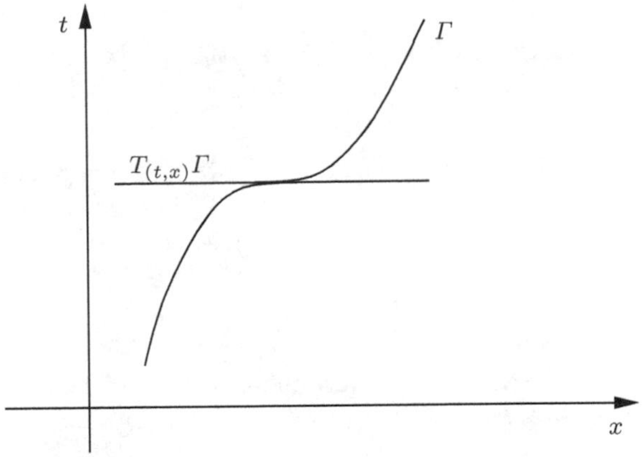

Fig. 2.13 This situation of a tangent space is excluded in the definition of a smoothly evolving hypersurface

In the above definition and also for $\Gamma(t)$ we allow for surfaces with boundary. We refer to Fig. 2.13 for an illustration what space-like means.

Let us fix $(t_0, x_0) \in \Gamma, x_0 \in \Gamma(t_0)$. Since Γ is a smooth hypersurface in $\mathbb{R} \times \mathbb{R}^n$ there exists an open neighborhood U of (t_0, x_0) and a smooth function $\phi : U \to \mathbb{R}$ with $(\phi_t, \nabla\phi) \neq 0$ in U and

$$U \cap \Gamma = \{(t, x) \in U \mid \phi(t, x) = 0\}.$$

Let $r > 0$ be chosen so small that $(t_0, x) \in U$ for $x \in B_r(x_0)$. Then

$$B_r(x_0) \cap \Gamma(t_0) = \{x \in B_r(x_0) \mid \phi(t_0, x) = 0\}.$$

Since $T_\Gamma(t_0, x_0) = [(\phi_t(t_0, x_0), \nabla\phi(t_0, x_0))]^\perp$ we infer from (ii) in the above definition that $\nabla\phi(t_0, x_0) \neq 0$. This implies that $\Gamma(t_0)$ is a smooth hypersurface in $\mathbb{R}^n$.

2.7 Normal Velocity and Normal Time Derivative

We now define the velocity of a surface. As a tangential movement on the evolving surface does not change the geometry of the surface the crucial quantity is the normal part of the velocity. In order to define the normal velocity we choose a normal vector field $\nu(t, \cdot)$ to $\Gamma(t)$.

Definition 2.7.1 Let $\Gamma = (\Gamma(t))_{t\in(0,T)}$ be a smoothly evolving family of hypersurfaces and $(t_0, x_0) \in \Gamma$.

(i) The normal velocity of Γ at (t_0, x_0) is defined as

$$V(t_0, x_0) = \nu(t_0, x_0) \cdot \frac{d\gamma}{dt}(t_0),$$

where $\gamma : (t_0 - \delta, t_0 + \delta) \to \mathbb{R}^n$ is a curve with $\gamma(t) \in \Gamma(t)$ and $\gamma(t_0) = x_0$.

(ii) Let $f : \Gamma \to \mathbb{R}$ be a smooth function. Then we define the normal time derivative of f at (t_0, x_0) as

$$\partial_t^\square f(t_0, x_0) = \frac{d}{dt} f(t, \tilde{\gamma}(t))_{|t=t_0},$$

where $\tilde{\gamma} : (t_0 - \delta, t_0 + \delta) \to \mathbb{R}^n$ is a curve with $\tilde{\gamma}(t) \in \Gamma(t)$, $\tilde{\gamma}(t_0) = x_0$ and $\frac{d\tilde{\gamma}}{dt}(t_0) = V(t_0, x_0)\nu(t_0, x_0)$.

Remark 2.7.2

(i) In Exercise 7.3 it will be shown that a curve γ as in (i) exists and that the definition of V does not depend on the choice of γ.

(ii) The fact that a curve $\tilde{\gamma}$ with the properties required in (ii) exists is shown in Exercise 7.4.

(iii) Since

$$\partial_t^\square f(t_0, x_0) = \nabla_\Gamma f(t_0, x_0) \cdot (1, V(t_0, x_0)\nu(t_0, x_0)),$$

the quantity $\partial_t^\square f(t_0, x_0)$ is independent of the choice of $\tilde{\gamma}$.

(iv) Assume that f is extended to a neighborhood of Γ. Then it holds

$$\partial_t^\square f(t_0, x_0) = \frac{\partial f}{\partial t}(t_0, x_0) + V(t_0, x_0)\frac{\partial f}{\partial \nu}(t_0, x_0).$$

2.8 Velocity Fields and Material Time Derivatives Induced by the Motion of Material Points

We now consider a family of smoothly evolving oriented hypersurfaces (with or without boundary) whose evolution is induced by the motion of material points. However, there is no need to think of real physical material points. We start with a smooth initial hypersurface $\Gamma(0) = \Gamma_0$, and assume that

$$\Gamma(t) = \{x \in \mathbb{R}^n \mid x = \Phi(t, p) \text{ for some } p \in \Gamma_0\}, \tag{2.18}$$

where $\Phi : [0, T] \times \Gamma_0 \to \mathbb{R}^n$ is smooth and $\Phi(t, \cdot)$ is a diffeomorphism for all $t \in [0, T]$. As in Definition 2.6.1 we set $\Gamma := \{(t, x) \in \mathbb{R} \times \mathbb{R}^n \mid t \in (0, T), x \in \Gamma(t)\}$.

Definition 2.8.1 Let Γ be as above and $(t_0, x_0) \in \Gamma$ with $x_0 = \Phi(t_0, p_0)$.

(i) We define the velocity field $v : \Gamma \to \mathbb{R}^n$ which is induced by Φ at (t_0, x_0) via

$$v(t_0, x_0) = \frac{d}{dt}\Phi(t, p_0)_{|t=t_0}.$$

(ii) Let $f : \Gamma \to \mathbb{R}$ be a smooth function. We call

$$D_t f(t_0, x_0) := \frac{d}{dt} f(t, \Phi(t, p_0))_{|t=t_0}$$

the material time derivative of f at (t_0, x_0).

Remark 2.8.2

(i) The quantity $D_t f$ depends on Φ.
(ii) Assume that f is extended to a neighborhood of Γ. Then it holds

$$D_t f(t, x) = \frac{\partial f}{\partial t}(t, x) + v(t, x) \cdot \nabla f(t, x), \qquad (t, x) \in \Gamma.$$

(iii) By $\Phi(t, .)$ boundary points are mapped to boundary points. Boundary points $\Phi(t, p)$ with $p \in \partial\Gamma_0$ hence travel with velocity $v(t, \Phi(t, p))$. We say the boundary evolves with velocity v.

Let us now assume that Γ_0 and hence the hypersurfaces $\Gamma(t)$ have an orientation given by a continuous normal field $\nu(t, \cdot)$. For $(t_0, x_0) \in \Gamma$ with $x_0 = \Phi(t_0, p_0)$ we set $\gamma(t) := \Phi(t, p_0)$. Clearly, $\gamma(t) \in \Gamma(t)$ and $\gamma(t_0) = x_0$ so that the definitions of the normal velocity V and the velocity field v imply that

$$V(t_0, x_0) = \nu(t_0, x_0) \cdot \gamma'(t_0) = \nu(t_0, x_0) \cdot v(t_0, x_0).$$

Therefore $v_\tau(t_0, x_0) := v(t_0, x_0) - V(t_0, x_0)\nu(t_0, x_0) \in T_{\Gamma(t_0)}(x_0)$ and we call $v_\tau : \Gamma \to \mathbb{R}^n$ the tangential velocity field induced by Φ.

Lemma 2.8.3

(i) For a velocity field $v = V\nu + v_\tau$ *on* Γ *we have*

$$\nabla_{\Gamma(t)} \cdot v = -V\kappa + \nabla_{\Gamma(t)} \cdot v_\tau.$$

(ii) For a function $f : \Gamma \to \mathbb{R}$ *it holds*

$$D_t f = \partial_t^{\Box} f + v_\tau \cdot \nabla_{\Gamma(t)} f.$$

Proof

(i) Proposition 2.5.3 implies that $\nabla_{\Gamma(t)} \cdot (V\nu) = V\nabla_{\Gamma(t)} \cdot \nu = -V\kappa$ and hence

$$\nabla_{\Gamma(t)} \cdot v = \nabla_{\Gamma(t)} \cdot (V\nu) + \nabla_{\Gamma(t)} \cdot v_\tau = -V\kappa + \nabla_{\Gamma(t)} \cdot v_\tau.$$

(ii) Let $\tilde{f}$ be an extension of f to a neighborhood of Γ. Then, the definitions of v and $D_t f$ imply that

$$D_t f = \frac{\partial \tilde{f}}{\partial t} + v \cdot \nabla \tilde{f} = \frac{\partial \tilde{f}}{\partial t} + V\nu \cdot \nabla \tilde{f} + v_\tau \cdot \nabla \tilde{f} = \partial_t^{\Box} f + v_\tau \cdot \nabla_{\Gamma(t)} f,$$

since $v_\tau \cdot \nu = 0$.

□

2.9 Jacobi's Formula for the Derivative of the Determinant

The following lemma will be crucial for a transport theorem which we will prove next.

Lemma 2.9.1 *Assume that* $t \mapsto A(t) \in \mathbb{R}^{d\times d}$ *is differentiable and* $A(t)$ *is invertible. Then it holds:*

$$\frac{d}{dt}\det A(t) = \operatorname{Tr}\left(A^{-1}(t)\frac{d}{dt}A(t)\right)\det A(t)\,.$$

For a proof see [57].

2.10 A Transport Theorem

Theorem 2.10.1 *Let* $\Gamma = (\Gamma(t))_{t\in(0,T)}$ *be a family of evolving hypersurfaces with a velocity field* $v = V\nu + v_\tau$ *on* Γ *as in Sect. 2.8. For a smooth* $f : \Gamma \to \mathbb{R}$ *it holds:*

$$\begin{aligned}\frac{d}{dt}\int_{\Gamma(t)} f\,d\mathcal{H}^{n-1} &= \int_{\Gamma(t)} (D_t f + f\nabla_{\Gamma(t)}\cdot v)\,d\mathcal{H}^{n-1}\\ &= \int_{\Gamma(t)} (D_t f + f\nabla_{\Gamma(t)}\cdot v_\tau - fV\kappa)\,d\mathcal{H}^{n-1}\\ &= \int_{\Gamma(t)} (\partial_t^{\square} f - fV\kappa)\,d\mathcal{H}^{n-1} + \int_{\partial\Gamma(t)} f\,\nu_{\partial\Gamma(t)}\cdot v_\tau\,d\mathcal{H}^{n-2}\,.\end{aligned}$$

Proof Let $\bar{F} : \Theta \to \mathbb{R}^n$ be a local parametrization of Γ_0 and define $F : J\times\Theta \to \mathbb{R}^n$ by $F(t,\theta) := \Phi(t,\bar{F}(\theta))$. Then $F(t,\cdot)$ is a local parametrization of $\Gamma(t)$ with

$$\partial_t F(t,\theta) = \partial_t\Phi(t,\bar{F}(\theta)) = v(t,\Phi(t,\bar{F}(\theta))) = v(t,F(t,\theta)) =: W(t,\theta), \quad (t,\theta)\in J\times\Theta.$$

In local coordinates the metric is given as

$$g_{ij}(t,\theta) = \partial_{\theta_i}F(t,\theta)\cdot\partial_{\theta_j}F(t,\theta), \quad i,j = 1,\ldots,n-1\,.$$

In addition, we define

$$G(t,\theta) = (g_{ij}(t,\theta))_{i,j=1,\dots,n-1}\,,$$
$$g(t,\theta) = \det G(t,\theta)\,,$$
$$G^{-1}(t,\theta) = (g^{ij}(t,\theta))_{i,j=1,\dots,n-1}\,.$$

With Jacobi's formula for the derivative of the determinant we obtain

$$\partial_t\sqrt{g} = \tfrac{1}{2}g^{-\frac{1}{2}}\operatorname{Tr}\left(G^{-1}\partial_t G\right)\det G = \tfrac{1}{2}\sqrt{g}\operatorname{Tr}\left(G^{-1}\partial_t G\right)\,.$$

The formula

$$\partial_t g_{ij} = \partial_t(\partial_{\theta_i}F\cdot\partial_{\theta_j}F) = \partial_{\theta_i}F\cdot\partial_{\theta_j}\partial_t F + \partial_{\theta_i}\partial_t F\cdot\partial_{\theta_j}F$$

now implies, using the symmetry of (g_{ij}) and the Einstein sum convention:

$$\begin{aligned}\partial_t\sqrt{g} &= \tfrac{1}{2}g^{\frac{1}{2}}g^{ij}(\partial_{\theta_j}F\cdot\partial_{\theta_i}\partial_t F + \partial_{\theta_j}\partial_t F\cdot\partial_{\theta_i}F)\\ &= g^{\frac{1}{2}}g^{ij}\partial_{\theta_i}F\cdot\partial_{\theta_j}W\,.\end{aligned}$$

Furthermore, if we set $\tilde{F}(t,\theta) := f(t,F(t,\theta)) = f(t,\Phi(t,\bar{F}(\theta))$ we obtain

$$\frac{\partial\tilde{F}}{\partial t}(t,\theta) = D_t f(t,\Phi(t,\bar{F}(\theta))) = D_t f(t,F(t,\theta)).$$

With this we obtain for a function which has his support in the image of the parametrization:

$$\begin{aligned}\frac{d}{dt}\int_\Gamma f\,d\mathcal{H}^{n-1} &= \frac{d}{dt}\int_\Theta \tilde{F}\sqrt{g}\,d\theta = \int_\Theta\left(\frac{\partial\tilde{F}}{\partial t}\sqrt{g} + \tilde{F}\frac{\partial\sqrt{g}}{\partial t}\right)d\theta\\ &= \int_\Theta\left(\frac{\partial\tilde{F}}{\partial t} + \tilde{F}g^{ij}\partial_{\theta_i}F\cdot\partial_{\theta_j}W\right)\sqrt{g}\,d\theta\\ &= \int_\Gamma (D_t f + f\nabla_\Gamma\cdot v)\,d\mathcal{H}^{n-1}\,.\end{aligned}$$

Using a partition of unity we obtain the first equality for general functions f. The second equality follows with the help of Lemma 2.8.3. The third equality then follows with the integration by parts formula on manifolds, see Proposition 2.5.1. □

2.11 Reynolds Transport Theorem

As a corollary of the above transport theorem we obtain the classical Reynolds theorem of continuum mechanics. Here we consider a family $(\Omega(t))_{t\in[0,T]}$ with $\Omega(t) \subset \mathbb{R}^n$ and interpret this as a hypersurface $\Omega(t) \times \{0\} \subset \mathbb{R}^{n+1}$ with normal velocity zero.

Theorem 2.11.1 *Let $(\Omega(t))_{t\in[0,T]}$ with $\Omega(t) \subset \mathbb{R}^n$ be a family of domains which evolve with a smooth velocity field v and let $f(t,x)$ be a smooth function defined on $(\Omega(t))_{t\in[0,T]}$. Then it holds*

$$\frac{d}{dt}\int_{\Omega(t)} f(t,x)dx = \int_{\Omega(t)} \left[\frac{\partial f}{\partial t}(t,x) + \nabla\cdot(f(t,x)v(t,x))\right] dx\,.$$

Proof The Transport Theorem 2.10.1 yields, since $V = 0$ and $v = v_\tau$,

$$\begin{aligned}\frac{d}{dt}\int_{\Omega(t)} f\,dx &= \int_{\Omega(t)} \left(\frac{\partial f}{\partial t} + \nabla f\cdot v + f\nabla\cdot v\right) dx \\ &= \int_{\Omega(t)} \left(\frac{\partial f}{\partial t} + \nabla\cdot(f(t,x)v(t,x))\right) dx\,.\end{aligned}$$

We also refer to [57] for a direct proof and precise assumptions on the smoothness of $(\Omega(t))_{t\in[0,T]}$ and f. □

Modeling

3

Abstract

In this chapter we derive important mathematical models involving interfaces. We start with one of the simplest evolution problems involving an interface: the mean curvature flow of grain boundaries. We will then discuss the gradient flow of curvature energies, the Stefan problem for melting and solidification, two-phase flows and phase field models. All these models will be analyzed mathematically in the later chapters.

In many physical systems an energy decreases in time. One of the simplest energies for a hypersurface $\hat{\Gamma}$ is the surface area of $\hat{\Gamma}$. We consider a smooth, compact, oriented hypersurface $\hat{\Gamma}$ in $\mathbb{R}^n$ without boundary and introduce the area functional

$$E(\hat{\Gamma}) := \mathscr{H}^{n-1}(\hat{\Gamma}) = \int_{\hat{\Gamma}} 1 \, d\mathscr{H}^{n-1} .$$

The goal now is to evolve $\hat{\Gamma}$ in such a way that the surface area decreases most rapidly. Roughly speaking this will be achieved by flowing $\hat{\Gamma}$ in the direction of the negative "gradient" of E. We will now introduce the concept of gradient flows first in a simple finite dimensional setting and then generalize the idea to the area functional.

E. Bänsch et al., *Interfaces: Modeling, Analysis, Numerics*,
Oberwolfach Seminars 51, https://doi.org/10.1007/978-3-031-35550-9_3

3.1 Gradient Flows

3.1.1 Gradient Flows in $\mathbb{R}^n$

For a sufficiently smooth function $\Phi : \mathbb{R}^n \to \mathbb{R}$ with derivative $D\Phi_{x_0}$ at the point $x_0 \in \mathbb{R}^n$ we define the gradient $\nabla \Phi(x_0) \in \mathbb{R}^n$ such that the following identity holds

$$D\Phi_{x_0}(v) = (\nabla \Phi(x_0)) \cdot v \qquad \text{for all } v \in \mathbb{R}^n .$$

Now $x : [0, T] \to \mathbb{R}^n$ is a solution of the gradient flow equation to Φ if

$$x'(t) = -\nabla \Phi(x(t)) \tag{3.1}$$

holds for all $t \in [0, T]$. In particular, we have

$$\begin{aligned} \frac{d}{dt}\Phi(x(t)) &= D\Phi_{x(t)}(x'(t)) \\ &= (\nabla\Phi(x(t))) \cdot x'(t) \\ &= -|\nabla \Phi(x(t))|^2 \leq 0 \end{aligned}$$

where $|.|$ denotes the Euclidean norm in $\mathbb{R}^n$. In particular, we obtain that $\Phi(x(t))$ can only decrease in time.

For any $y : [0, T] \to \mathbb{R}^n$ with $|y'(0)| = |\nabla \Phi(x(0))|$ and $y(0) = x(0)$ we have, using the Cauchy–Schwarz inequality,

$$\begin{aligned} \frac{d}{dt}\Phi(y(t))_{|t=0} &= (\nabla \Phi(y(0))) \cdot y'(0) \\ &= (\nabla \Phi(x(0))) \cdot y'(0) \\ &\geq -|\nabla \Phi(x(0))|^2 \end{aligned}$$

with an equality if and only if

$$y'(0) = -\nabla \Phi(x(0)) .$$

This shows that among all possible directions, the direction $-\nabla \Phi(x(0))$ decreases Φ most efficiently.

3.1.2 Minimizing Movements for Gradient Flows

For gradient flows one can define an energy-driven implicit time discretization for the approximation of solutions to gradient flows. The idea has been introduced by De Giorgi. For a detailed discussion of gradient flows and minimizing movements we refer to the book by Ambrosio, Gigli and Savaré [8]. We here only describe the idea in the simple case of gradient flows in $\mathbb{R}^n$. However in Sect. 5.6 we will discuss a more complex situation in which the idea of minimizing movements for gradient flows is used for interface problems. In addition, the idea can also be used for the phase field models discussed in Sect. 3.8.

We now consider the gradient flow (3.1). Given x_0 and $\delta > 0$ we define recursively x_k as a minimizer of

$$\min F_k^\delta(x) := \left\{ \Phi(x) + \frac{1}{2\delta}|x - x_{k-1}|^2 \right\}.$$

We now interpret x_k as the solution at time $k\delta$ and extend the values to a function u^δ on $[0, \infty)$ as a piece-wise constant function such that u^δ is constant on $[k\delta, (k+1)\delta)$ for $k = 0, 1, 2, 3, \ldots$. Under appropriate assumptions the minimizer x_k fulfills $\nabla F_k^\delta(x_k) = 0$ which implies

$$\frac{x_k - x_{k-1}}{\delta} = -\nabla\Phi(x_k) \tag{3.2}$$

which can be interpreted as an implicit time discretization of (3.1). Equation (3.2) implies

$$\frac{u^\delta(t) - u^\delta(t-\delta)}{\delta} = -\nabla\Phi(u^\delta(t)).$$

Assuming that u^δ has a limit u as δ tends to 0 (in an appropriate sense) we obtain by formally passing to the limit in the above identity that

$$u'(t) = -\nabla\Phi(u(t)).$$

As it is in general very easy to obtain, under appropriate assumptions, solutions to variational problems the minimizing movement approach is very attractive both from an analytical as well as from a numerical point of view.

3.2 First Variation of Area

In order to define the gradient of

$$E(\hat{\Gamma}) := \mathcal{H}^{n-1}(\hat{\Gamma}) = \int_{\hat{\Gamma}} 1\, d\mathcal{H}^{n-1}$$

we first of all need to determine the first variation (the "derivative") of the area functional. We always consider a smooth, compact, oriented hypersurface $\hat{\Gamma}$ in $\mathbb{R}^n$ without boundary.

In order to compute a directional derivative of E we need to embed $\hat{\Gamma}$ in a one-parameter family of surfaces. This will be achieved with the help of a smooth vector field $\zeta : \mathbb{R}^n \to \mathbb{R}^n$. We define

$$\Gamma(t) := \{x + t\zeta(x) \mid x \in \hat{\Gamma}\}, \quad t \in \mathbb{R} \tag{3.3}$$

and observe that $\Gamma(0) = \hat{\Gamma}$. Using the Transport Theorem 2.10.1 we obtain (note that $v = \zeta$ on $\hat{\Gamma}$)

$$\frac{d}{dt} E(\Gamma(t))_{|t=0} = \int_{\hat{\Gamma}} (\nabla_\Gamma \cdot v_\tau - V\kappa) d\mathcal{H}^{n-1}$$

with $V = \zeta \cdot \nu$ and $v_\tau = \zeta - (\zeta \cdot \nu)\nu$. As $\hat{\Gamma}$ has no boundary, Corollary 2.5.2(a) gives

$$\frac{d}{dt} E(\Gamma(t))_{|t=0} = -\int_{\hat{\Gamma}} V\kappa d\mathcal{H}^{n-1}.$$

3.3 Mean Curvature Flow as a Gradient Flow of the Area Functional

We formally endow the space $\mathcal{M}$ of all closed, oriented hypersurfaces $\hat{\Gamma}$ in $\mathbb{R}^n$ with a tangent space which consists of all possible normal velocities, i.e., we set

$$T_{\mathcal{M}}(\hat{\Gamma}) = \{V : \hat{\Gamma} \to \mathbb{R}\}.$$

We refer to the book of Prüss and Simonett [129] for a rigorous discussion. A function $V : \hat{\Gamma} \to \mathbb{R}$ arises as a "tangent" vector, i.e. as a differential of a curve in $\mathcal{M}$, if we consider a vector field $\zeta : \mathbb{R}^n \to \mathbb{R}^n$ such that $\zeta \cdot \nu = V$ on $\hat{\Gamma}$ and define $\Gamma(t)$ as in (3.3). One natural choice of an inner product on $T_{\mathcal{M}}(\hat{\Gamma})$ is given by

$$\langle v_1, v_2 \rangle_{L^2} = \int_{\hat{\Gamma}} v_1 v_2\, d\mathcal{H}^{n-1} \quad \text{for all } v_1, v_2 \in T_{\mathcal{M}}(\hat{\Gamma}).$$

Now the gradient $\mathrm{grad}_{\mathscr{M}} E$ of E needs to fulfill

$$\langle \mathrm{grad}_{\mathscr{M}} E, V\rangle_{L^2} = \frac{d}{dt} E(\Gamma(t))_{|t=0} = -\int_{\hat{\Gamma}} \kappa V d\mathscr{H}^{n-1}$$

for all $V : \hat{\Gamma} \to \mathbb{R}$. We hence obtain

$$\mathrm{grad}_{\mathscr{M}} E = -\kappa$$

and the gradient flow of the area functional E is the mean curvature flow

$$V = \kappa\,.$$

More precisely, we say that a smooth one–parameter family $(\Gamma(t))_{t\geq 0}$ of hypersurfaces in $\mathbb{R}^n$ solves $V = \kappa$ if for a local parametrization $F(t, p)$, $p \in U$, $U \subset \mathbb{R}^{n-1}$ open, it holds that

$$\partial_t F \cdot \nu = \kappa\,.$$

In particular, we obtain

$$\frac{d}{dt}\mathscr{H}^{n-1}(\Gamma(t)) = -\int_{\Gamma(t)} \kappa^2 \leq 0\,.$$

For more information on mean curvature flow we refer to the articles by Ecker [59] and the books [32, 58, 94, 116].

3.4 Anisotropic Energies and Their Gradient Flows

In the introduction we mentioned crystal growth as one important situation in which interfaces appear. We also already introduced interfacial energies given as the total surface area of the interface. However, often the interfacial energy is anisotropic meaning that the density of the interfacial energy depends on the local orientation of the interface. This happens for example for crystals as depending on the crystal structure it might be energetically favorable to have an interface in certain directions, see, e.g. Fig. 1.1. The local orientation of an interface can be identified with the normal ν to the interface. We will now choose a positive function $\gamma : \mathbb{S}^{n-1} \to (0, \infty)$, $\mathbb{S}^{n-1}$ being the unit sphere, such that $\gamma(\nu)$ gives the local interfacial energy density. A typical anisotropic energy has the form

$$E_\gamma(\hat{\Gamma}) = \int_{\hat{\Gamma}} \gamma(\nu)\, d\mathscr{H}^{n-1}(x) \tag{3.4}$$

where $\hat{\Gamma}$ is a closed orientable hypersurface. Here γ describes how "expensive" it is to have an interface with normal ν. It will be convenient to extend γ from the unit sphere $\mathbb{S}^{n-1}$ to $\mathbb{R}^n$ via

$$\gamma(\lambda p) = \lambda\gamma(p) \qquad \text{for all} \qquad p \in \mathbb{S}^{n-1}, \quad \lambda > 0, \tag{3.5}$$

so that γ is positively homogeneous of degree one. We assume from now on that $\gamma \in C^2(\mathbb{R}^n \setminus \{0\})$. Under this assumption we can differentiate the identity (3.5) with respect to λ to obtain

$$\gamma'(p) \cdot p = \gamma(p) \quad \text{for all } p \in \mathbb{R}^n \setminus \{0\}, \tag{3.6}$$

where γ' is the gradient of γ. In the isotropic case, $\gamma(p) = |p|$, and so $\gamma'(p) = \frac{p}{|p|}$. We refer to [24, 82, 146] for more details on anisotropic energies in materials science and geometry.

We can formulate an isoperimetric problem for the surface energy E_γ. The problem is to find a shape which minimizes E_γ under all shapes with a given enclosed volume. In order to do so, one defines the dual function

$$\gamma^*(q) = \sup_{p\in\mathbb{R}^n\setminus\{0\}} \frac{p \cdot q}{\gamma(p)} \quad \text{for all } q \in \mathbb{R}^n .$$

Then the solution of the isoperimetric problem is, up to a scaling, the Wulff shape

$$\mathscr{W} = \{q \in \mathbb{R}^n : \gamma^*(q) \leq 1\},$$

see [82] and the references therein for more details. This is the 1-ball of γ^* and we also define the 1-ball of γ

$$\mathscr{F} = \{p \in \mathbb{R}^n : \gamma(p) \leq 1\},$$

which is called Frank diagram. We refer to Figs. 3.1, 3.2 and 3.3 for examples.

We now want to compute the first variation of the energy functional E_γ. To do so we first need to be able to compute for an evolving hypersurface the time derivative of the normal. This will be done in the following lemma.

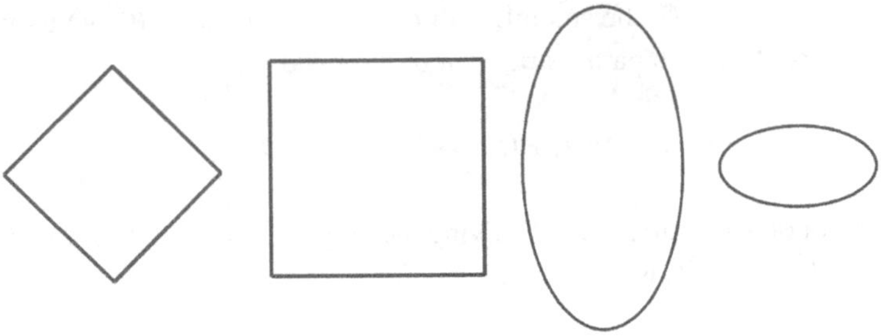

Fig. 3.1 Frank diagram and Wulff shape in $\mathbb{R}^2$ for the l^1-norm, $\gamma(\vec{p}) = \sum_{i=1}^{2} |p_i|$, left, and for the weighted norm $\gamma(\vec{p}) = (\vec{p} \cdot G\,\vec{p})^{\frac{1}{2}}, G = \frac{1}{4}\begin{pmatrix} 4 & 0 \\ 0 & 1 \end{pmatrix}$, right

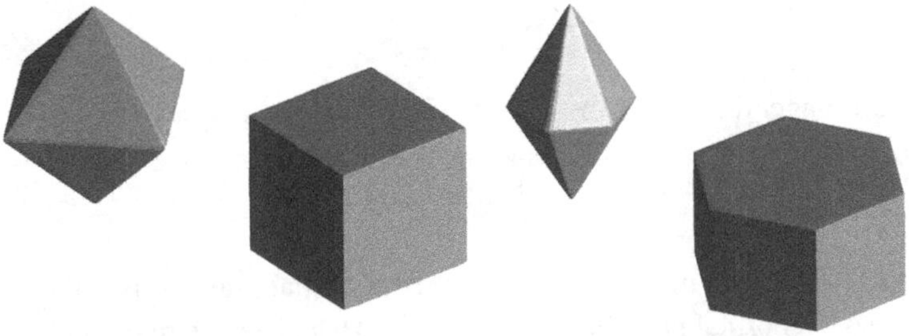

Fig. 3.2 Frank diagrams and Wulff shapes for different choices of the anisotropy, see [23] for details

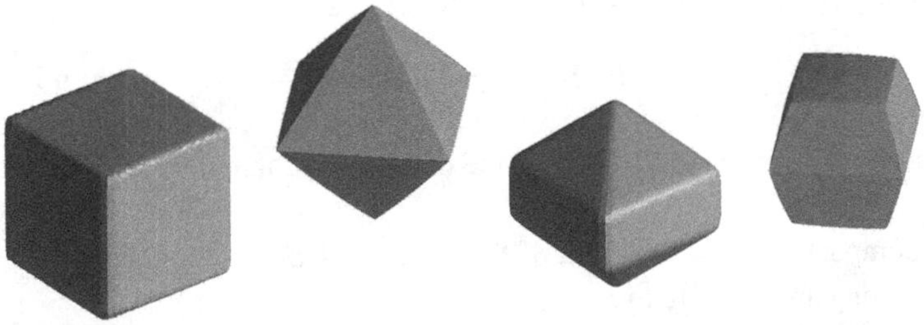

Fig. 3.3 Frank diagrams and Wulff shapes for different choices of the anisotropy, see [23] for details

Lemma 3.4.1 *Let $(\Gamma(t))_{t\in(0,T)}$ be a smoothly evolving family of orientable hypersurfaces with velocity v, normal velocity V and normal ν. Then it holds for the material time derivative*

$$D_t \nu = -(\nabla_\Gamma v)^T \nu \tag{3.7}$$

and for the normal time derivative we have

$$\partial_t^\square \nu = -\nabla_\Gamma V. \tag{3.8}$$

Proof Let $F : J \times U \to \mathbb{R}^n$ be smooth with $U \subset \mathbb{R}^{n-1}$ and $J \subset (0, T)$ an interval such that $F(t, .)$ is a local smooth parametrization of $\Gamma(t)$ and

$$F_t(t, \theta) = v(t, F(t, \theta)), \text{ for all } \theta \in U, \ t \in J,$$

see also the proof of Theorem 2.10.1. Defining the tangential vectors τ_i via $\tau_i(t, F(t, \theta)) = \partial_i F(t, \theta), \ i = 1, ..., n-1$, we now compute, using that $\nu \cdot \tau_i = 0$,

$$\begin{aligned}(D_t \nu) \cdot \tau_i &= -\nu \cdot D_t \tau_i = -\nu \cdot \partial_t \partial_i F \\ &= -\nu \cdot \partial_i \partial_t F = -\nu \cdot ((\nabla_\Gamma v) \, \tau_i) \\ &= -\left((\nabla_\Gamma v)^T \nu\right) \cdot \tau_i .\end{aligned}$$

This shows the tangential part of the identity (3.7). We now notice that

$$D_t \nu \cdot \nu = \tfrac{1}{2} D_t |\nu|^2 = \tfrac{1}{2} D_t 1 = 0.$$

As $\nu \cdot ((\nabla_\Gamma v)^T \nu) = ((\nabla_\Gamma v)\nu) \cdot \nu$ and since the tangential gradient is orthogonal to ν we obtain that also the normal parts in (3.7) are equal. This gives the first identity.

The second identity follows from

$$\begin{aligned}\partial_t^\square \nu &= D_t \nu - (\nabla_\Gamma \nu) \, v_\tau = D_t \nu - (\nabla_\Gamma \nu)^T v_\tau = -(\nabla_\Gamma v)^T \nu - (\nabla_\Gamma \nu)^T v_\tau \\ &= -(\nabla_\Gamma v)^T \nu - (\nabla_\Gamma \nu)^T v = -\nabla_\Gamma (v \cdot \nu) = -\nabla_\Gamma V \qquad \text{on } \Gamma(t),\end{aligned}$$

where we used the formula (3.7) for the time derivative of the normal. □

Before we compute the first variation of the functional E_γ we first show how the energy E_γ changes along an evolving hypersurface.

Theorem 3.4.2 *Let* $(\Gamma(t))_{t \in (0,T)}$ *be a smoothly evolving family of orientable hypersurfaces without boundary and let* ν *be the normal vector field. Then we obtain*

$$\frac{d}{dt} E_\gamma(\Gamma(t)) = -\int_{\Gamma(t)} \kappa_\gamma V \, d\mathcal{H}^{n-1} \tag{3.9}$$

where $\kappa_\gamma = -\nabla_\Gamma \cdot \gamma'(\nu)$ *is called anisotropic mean curvature or weighted mean curvature.*

Proof Using the Transport Theorem 2.10.1, the identity (3.6), Lemma 3.4.1 and the integration by parts formula on hypersurfaces from Proposition 2.5.1 we compute

$$\begin{aligned}\frac{d}{dt}E_\gamma(\Gamma(t)) &= \int_{\Gamma(t)} \partial_t^\square \gamma(\nu) - V\kappa\gamma(\nu) d\mathcal{H}^{n-1} \\ &= \int_{\Gamma(t)} \gamma'(\nu)\partial_t^\square \nu - \gamma'(\nu)\cdot \nu\, V\kappa\, d\mathcal{H}^{n-1} \\ &= \int_{\Gamma(t)} (-\gamma'(\nu))\cdot \nabla_\Gamma V + \nabla_\Gamma \cdot (\gamma'(\nu)V)\, d\mathcal{H}^{n-1} \\ &= \int_{\Gamma(t)} V\nabla_\Gamma \cdot \gamma'(\nu)\, d\mathcal{H}^{n-1} \\ &= -\int_{\Gamma(t)} \kappa_\gamma V\, d\mathcal{H}^{n-1}.\end{aligned}$$

□

We remark that κ_γ is the mean curvature in the isotropic case which is given as $\gamma(\nu) = |\nu|$. In this case E_γ is the surface area and we have $-\nabla_\Gamma \cdot \gamma'(\nu) = -\nabla_\Gamma \cdot \nu = \kappa$.

In order to compute a directional derivative of E_γ we embed a hypersurface $\hat{\Gamma}$ in a one-parameter family of surfaces. This will be achieved as in Sect. 3.2 with the help of a smooth vector field $\zeta : \mathbb{R}^n \to \mathbb{R}^n$. We define

$$\Gamma(t) := \{x + t\zeta(x) \mid x \in \hat{\Gamma}\}, \quad t \in \mathbb{R}. \tag{3.10}$$

Using Theorem 3.4.2 we obtain that

$$\frac{d}{dt}E_\gamma(\Gamma(t))_{|t=0} = -\int_{\hat{\Gamma}} \kappa_\gamma V\, d\mathcal{H}^{n-1}, \tag{3.11}$$

where $V = \zeta \cdot \nu$. Analogously as in Sect. 3.3 we obtain from Theorem 3.4.2 that

$$V = \kappa_\gamma \tag{3.12}$$

is the gradient flow of E_γ.

3.5 The Gradient Flow of the Willmore Functional

In this section we want to compute the L^2-gradient flow of the Willmore functional

$$E_W(\hat{\Gamma}) = \frac{1}{2}\int_{\hat{\Gamma}} \kappa^2(x)\, d\mathcal{H}^{n-1}(x). \tag{3.13}$$

In order to do so we need some geometric results.

Lemma 3.5.1

(i) Let $\hat{\Gamma}$ be an orientable C^3-hypersurface with normal field ν. Then it holds that

$$\nabla_\Gamma \kappa = -\Delta_\Gamma \nu - |\nabla_\Gamma \nu|^2 \nu \qquad \text{on} \quad \hat{\Gamma}. \tag{3.14}$$

(ii) Let $(\Gamma(t))_{t\in(0,T)}$ be a family of evolving hypersurfaces with normal vector field $\nu(t,\cdot)$ and mean curvature $\kappa(t,\cdot)$. Then it holds

$$D_t\kappa = \Delta_\Gamma V + V|\nabla_\Gamma \nu|^2 + v_\tau \cdot \nabla_\Gamma \kappa \qquad \text{on } \Gamma(t), \tag{3.15}$$

$$\partial_t^\square \kappa = \Delta_\Gamma V + V|\nabla_\Gamma \nu|^2 \qquad \text{on } \Gamma(t). \tag{3.16}$$

Here $|A| = \sqrt{tr(A^T A)}$ is the Frobenius norm of a matrix A.

The proof is left to an exercise, see Chap. 7 for some hints. We are now in a position to compute the first variation of E_W.

Theorem 3.5.2 *Let $\Gamma = (\Gamma(t))_{t\in(0,T)}$ be a family of smoothly evolving closed and orientable hypersurfaces. Then it holds that*

$$\frac{d}{dt}E_W(\Gamma(t)) = \int_{\Gamma(t)} (\Delta_\Gamma\kappa + \kappa|\nabla_\Gamma \nu|^2 - \frac{1}{2}\kappa^3) V\, d\mathcal{H}^{n-1}. \tag{3.17}$$

Proof Using the Transport Theorem 2.10.1, Lemma 3.5.1 and the integration by parts formula on hypersurfaces from Proposition 2.5.1 we compute

$$\begin{aligned}\frac{d}{dt}E_W(\Gamma(t)) &= \int_{\Gamma(t)} \kappa(\partial_t^{\square}\kappa - \frac{1}{2}\kappa^2 V)\, d\mathcal{H}^{n-1} \\ &= \int_{\Gamma} \kappa(\Delta_\Gamma V + V|\nabla_\Gamma \nu|^2 - \frac{1}{2}\kappa^2 V)\, d\mathcal{H}^{n-1} \\ &= \int_{\Gamma} (\Delta_\Gamma \kappa + |\nabla_\Gamma \nu|^2 \kappa - \frac{1}{2}\kappa^3) V \, d\mathcal{H}^{n-1}.\end{aligned}$$

Here we used that the surfaces $\Gamma(t)$ have no boundary. □

With the help of the variations

$$\Gamma(t) := \{x + t\zeta(x) \mid x \in \hat{\Gamma}\}, \quad t \in \mathbb{R} \tag{3.18}$$

we obtain similarly as in the sections before

$$V = -\Delta_\Gamma \kappa - |\nabla_\Gamma \nu|^2 \kappa + \frac{1}{2}\kappa^3 \tag{3.19}$$

as the L^2-gradient flow of E_W which is known as Willmore flow.

3.6 The Stefan Problem

Interfaces in the natural sciences typically appear when the phase of a physical state changes. An example is melting and solidification of water or of a metal. This phenomenon is described by the Stefan problem. In the case of melting and solidification different phases differ in the constitutive relation between internal energy and temperature. We will now discuss this aspect for the solid-liquid phase transition with the help of a simple constitutive relation. For the internal energy we set

$$u(\theta) = \begin{cases} c_V\theta & \text{in the solid phase,} \\ c_V\theta + L & \text{in the liquid phase,} \end{cases} \tag{3.20}$$

where θ is the temperature, u is the internal energy, c_V is the specific heat and L is the latent heat. This constitutive relation reflects the following experimentally verified fact. There are temperatures at which energy can be supplied to a system without an increase of the temperature. At such a temperature a phase transition occurs. The energy needed to change the phase is called *latent heat*. In the above constitutive relation the latent heat is

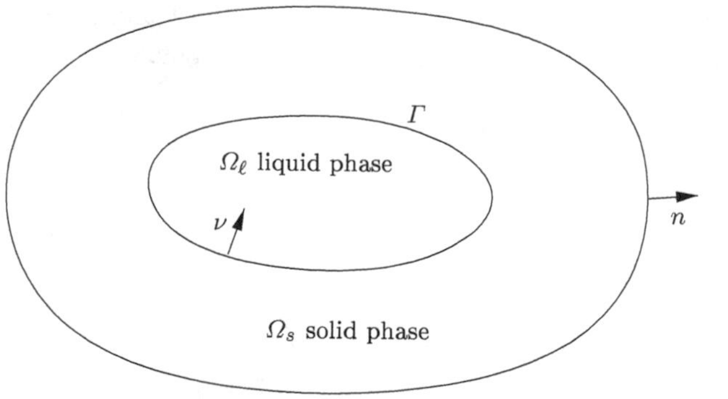

Fig. 3.4 Illustration of the geometry in the Stefan problem

called L. At the transition from solid to liquid at the melting temperature the latent heat is needed to enable the body to melt and hence to change its phase. This effect is used when you cool a liquid with ice cubes as the melting ice cubes withdraws heat from the surrounding.

We now consider a domain $\Omega \subset \mathbb{R}^n$ which is at each time $t \in [0, T]$ divided into a solid phase $\Omega_s(t)$, a liquid phase $\Omega_l(t)$ and an interface $\Gamma(t)$, see Fig. 3.4. For all domains $U \subset \Omega$ the energy conservation law in this case is given as

$$\frac{d}{dt}\int_U \rho u dx + \int_{\partial U} q \cdot n \, d\mathcal{H}^{n-1} = 0,$$

where ρ is the constant mass density, q is the heat flux and n is the outer unit normal to ∂U. This identity has to hold for all smooth, open $U \subset \Omega$.

3.6.1 Governing Equations in the Bulk

If U lies completely in the solid or in the liquid and if u and q are smooth we obtain with the help of the divergence theorem

$$\int_U (\partial_t(\rho u) + \nabla \cdot q) dx = 0$$

and as this identity has to hold for all open and smooth U the fundamental theorem of the calculus of variations gives

$$\partial_t(\rho u) + \nabla \cdot q = 0$$

pointwise in the solid and liquid phases. In the following we always assume that

$$q = -\lambda \nabla \theta \,, \tag{3.21}$$

i.e., there is a heat flux from regions of high temperature to regions of low temperature. This is in accordance with every day experience and experimental facts. In addition, we assume for simplicity that λ and the specific heat c_V are constant and are in particular the same in both phases. This implies that the heat equation

$$\rho c_V \partial_t \theta = \lambda \Delta \theta \tag{3.22}$$

is fulfilled in both phases.

Which equations should be postulated at the phase boundary?

Equation 1: The temperature θ is continuous across the phase boundary.

In most cases this assumption holds and is motivated from experiments and thermodynamics. This is also the working principle of a thermometer.

Another equation follows from the fact that ρu jumps at the interface. In order to deal with this we make the following assumption.

Assumption: The latent heat L is constant and in particular does not depend on θ.

The internal energy and hence the integrand ρu are discontinuous at the phase boundary, cf. Eq. (3.20), and in the term $\frac{d}{dt}\int_\Omega \rho u\, dx$ we cannot interchange the time derivative and the integral. In order to interchange time derivative and integral we need the following form of the transport theorem.

3.6.2 Another Transport Theorem

Theorem 3.6.1 (Transport Theorem) *Let $Q = (0,T) \times \Omega$ where $T > 0$ and Ω is a bounded Lipschitz domain in $\mathbb{R}^n$. We further assume that Γ in Q is a smoothly evolving hypersurface such that for all $t \in (0,T)$ the hypersurfaces $\Gamma(t)$ are compact subsets of Ω which separate domains $\Omega_s(t)$ and $\Omega_l(t)$. Furthermore, we define domains $Q_s = \{(t,x) \mid \text{with } t \in (0,T) \text{ and } x \in \Omega_s(t)\}$ and $Q_\ell = \{(t,x) \mid \text{with } t \in (0,T) \text{ and } x \in \Omega_\ell(t)\}$ and choose the unit normal ν to $\Gamma(t)$ to be pointing into the set $\Omega_\ell(t)$. In addition, let $u : Q \to \mathbb{R}$ be such that $u_{|Q_\ell}$ and $u_{|Q_s}$ have extensions on $\overline{Q}_\ell$ and $\overline{Q}_s$ which are continuously differentiable. Under*

(continued)

these assumptions it holds that

$$\frac{d}{dt}\int_{\Omega} u(t,x)\,dx = \int_{\Omega_\ell(t)} \partial_t u(t,x)\,dx + \int_{\Omega_s(t)} \partial_t u(t,x)\,dx - \int_{\Gamma(t)} [u]_s^\ell V \, d\mathcal{H}^{n-1}. \tag{3.23}$$

Here we set for $x \in \Gamma(t)$

$$[u]_s^\ell(t,x) := \lim_{\substack{y\to x\\ y\in\Omega_\ell(t)}} u(t,y) - \lim_{\substack{y\to x\\ y\in\Omega_s(t)}} u(t,y).$$

This theorem can be shown with the help of the Reynolds Transport Theorem 2.11.1 from Chap. 2 (apply it for the individual phases separately) or with the divergence theorem, see [57, Section 7.3].

3.6.3 Governing Equations on the Interface

We now consider a domain $U \subset \Omega$ with smooth boundary. From the identity

$$\frac{d}{dt}\int_U \rho u \, dx + \int_{\partial U} q \cdot n \, d\mathcal{H}^{n-1} = 0$$

it follows with the help of the transport theorem, the divergence theorem, and the energy conservation in the solid and liquid phase

$$\begin{aligned} 0 &= \int_U (\rho\, \partial_t u + \nabla \cdot q)\,dx + \int_{U\cap\Gamma(t)} \big(-\rho[u]_s^\ell V + [q]_s^\ell \cdot \nu\big) d\mathcal{H}^{n-1} \\ &= \int_{U\cap\Gamma(t)} \big(-\rho[u]_s^\ell V + [q]_s^\ell \cdot \nu\big) d\mathcal{H}^{n-1}. \end{aligned}$$

The above identity hence holds for arbitrary relatively open subsets of $\Gamma(t)$ and we obtain the following local form of the energy conservation law on the free boundary

$$\rho[u]_s^\ell V = [q]_s^\ell \cdot \nu \quad \text{on} \quad \Gamma(t). \tag{3.24}$$

This condition corresponds to the Rankine–Hugoniot condition for hyperbolic conservation laws and in the context of phase transitions it is called the *Stefan condition*. Using the definition of u we deduce

$$[u]_s^\ell = (c_V\theta + L - c_V\theta) = L\,.$$

Now, let q_s and q_ℓ be the fluxes in $\overline{\Omega}_s$ and $\overline{\Omega}_\ell$, respectively, and $\nu_s = \nu$, $\nu_\ell = -\nu$ the outer unit normals to Ω_s and Ω_ℓ, respectively. Altogether we obtain

$$q_\ell \cdot \nu_\ell + q_s \cdot \nu_s = -\rho L V\,.$$

In particular this means that if $q_\ell \cdot \nu_\ell + q_s \cdot \nu_s$ is positive the solid is melting. The expression on the left-hand side gives the total heat supplied by the two phases at the phase boundary. The more heat enters the phase boundary the quicker the melting process will be and the velocity of the phase boundary is proportional to the total energy flux into the interface. The heat entering the phase boundary yields the latent heat that is needed for the formation of the new liquid phase region. If the total energy flux into the phase boundary is negative heat is withdrawn from the phase boundary and the liquid solidifies. This implies that the latent heat is set free which compensates the negative total energy flux. So far we have two conditions at the interface $\Gamma(t)$ (for simplicity we set $\lambda = \rho = c_V = L = 1$)

- θ is continuous,
- $V + [\nabla\theta]_s^l \cdot \nu = 0$.

However, the heat equations in the liquid and solid phases *each* need one boundary condition and the normal velocity is another degree of freedom at the interface. So far we have two equations and there are different possibilities to fix the remaining degree of freedom.

Possibility I: $\theta = \theta_M$ with θ_M being the melting temperature.

In this case we obtain the following problem: Find a liquid phase Q_ℓ, a solid phase Q_s, a free boundary Γ, separating the two phases, and a temperature $\theta : Q \to \mathbb{R}$ such that

$$\partial_t\theta - \Delta\theta = 0 \quad \text{in } Q_s \cup Q_\ell\,, \tag{3.25}$$

$$V + [\nabla\theta]_s^\ell \cdot \nu = 0 \quad \text{on } \Gamma\,, \tag{3.26}$$

$$\theta = \theta_M \quad \text{on } \Gamma\,. \tag{3.27}$$

In addition, we need to specify initial conditions for θ and Γ, and for θ we also require boundary conditions on $\partial\Omega$. In the following we set the heat flux through the boundary to zero, i.e.,

$$-\nabla\theta \cdot n = 0 \quad \text{on } (0, T) \times \partial\Omega\,.$$

Considering the condition (3.26) we notice that $\nabla\theta$ has to jump across Γ whenever $V \neq 0$ holds. This is the case if the phase boundary is moving in time.

The free boundary problem (3.25)–(3.27) is the classical Stefan problem for melting and solidification. In this model, as stated above, one requires that in the solid phase $\theta < \theta_M$ holds and that in the liquid phase $\theta > \theta_M$ is true. In this case we can write the Stefan problem in a compact form which is called the *enthalpy formulation*

$$\partial_t \left(\theta + \chi_{\{\theta > \theta_M\}}\right) = \Delta\theta \,. \tag{3.28}$$

The expression $\chi_{\{\theta>\theta_M\}}$ is the characteristic function of the set $\{(t,x) \mid \theta(t,x) > \theta_M\}$, i.e., $\chi_{\{\theta>\theta_M\}}$ is 1 in the liquid phase and 0 in the solid phase. This formulation follows from the identity $u(\theta) = \theta + \chi_{\{\theta>\theta_M\}}$. Due to the fact that $\chi_{\{\theta>\theta_M\}}$ is not differentiable it is not possible to interpret Eq. (3.28) in a classical sense. Hence we interpret the identity (3.28) in a distributional sense, i.e., for all $\zeta \in C_0^\infty(Q)$, $Q = (0,T) \times \Omega$ we require

$$\int_Q \left((\theta + \chi_{\{\theta>\theta_M\}})\partial_t\zeta + \theta\Delta\zeta \right) dx\, dt = 0 \,. \tag{3.29}$$

We now seek a function $\theta(t,x)$ which fulfills (3.28) in a distributional sense. Having determined θ we obtain the liquid and the solid phase a posteriori as the sets $\Omega_\ell(t) = \{x \mid \theta(t,x) > \theta_M\}$ and $\Omega_s(t) = \{x \mid \theta(t,x) < \theta_M\}$. The phase boundary is given as $\Gamma(t) = \{x \mid \theta(t,x) = \theta_M\}$. However, there are situations in which $\Gamma(t)$ is not a hypersurface anymore and has a nonempty interior. In such a situation one says that a "mushy region" has formed.

Introducing the quantity

$$e = \begin{cases} \theta & \text{for } \theta \le \theta_M \,, \\ (\theta + 1) & \text{for } \theta > \theta_M \end{cases}$$

and defining

$$\beta(e) := \begin{cases} e & \text{for } e < \theta_M \,, \\ \theta_M & \text{for } \theta_M \le e \le (\theta_M + 1) \,, \\ (e - 1) & \text{for } e > (\theta_M + 1) \,, \end{cases}$$

we can formally rewrite Eq. (3.28) as

$$\partial_t e = \Delta\beta(e) \,. \tag{3.30}$$

Here it is important to notice that β is not strictly monotonically increasing. As a consequence the formulation (3.30) leads to a degenerate parabolic equation. For the numerical approximation of solutions to the Stefan problem the formulation (3.30) on the other hand has many advantages. In particular, a simple explicit time discretization can be used to construct approximate solutions.

In Exercise 7.9 you show that the distributional formulation (3.29) leads to the Stefan condition

$$V = -[\nabla\theta]_s^l \cdot \nu$$

which has to hold on the interface Γ.

Possibility II: So far we assumed that the liquid phase is characterized by $\theta > \theta_M$ and that the solid phase is characterized by $\theta < \theta_M$. In fact, it is possible that liquids are undercooled and that solids are superheated. It may happen, for instance, that liquids remain in the liquid phase even when the temperature in the liquid is below the melting temperature. We now set $u = \theta - \theta_M$. It is noticed in physics that the surface energy and the velocity of the interface have an effect on the temperature at the interface. In fact on Γ one considers

$$\beta V = \gamma\kappa - u \tag{3.31}$$

with $\gamma, \beta \geq 0$ instead of $u = 0$. The term βV takes kinetic undercooling into account and the term $\gamma\kappa$ allows for curvature undercooling. In fact due to both terms, e.g. water can freeze well beneath the melting temperature.

In fact Eq. (3.31) can be interpreted as a forced mean curvature flow equation where the right hand side u has to be computed as a solution of heat equations in solid and liquid.

Mushy regions as in the case of Possibility I do not appear in this case. We also remark that the condition $\theta = \theta_M$ in the case of undercooling leads to very unstable situations. One observes in the case without capillary term $\gamma\kappa$ and strong undercooling very unstable phase boundaries. Prescribing Eq. (3.31) implies that the formation of new surface costs energy and one obtains that the capillary term has a stabilizing effect such that small wavelengths in perturbations of the interface are damped.

The fact that strongly undercooled fluids have very unstable phase boundaries yields very bifurcated phase boundaries. Many solidification fronts lead to dendritic (tree-like) structures. Variants of the Stefan problem above are used to explain the diverse patterns observed in snow crystal growth (see, e.g., Libbrecht [109], Barrett, Garcke, Nürnberg [18] and Fig. 1.2).

3.7 Mathematical Modeling of Two-Phase Flows

In this section we derive a mathematical model to describe two phase flow. The situation is depicted in Fig. 3.5. One considers the flow of two incompressible, immiscible fluids (for instance oil and water or a gas bubble rising in water) separated by a sharp and smooth

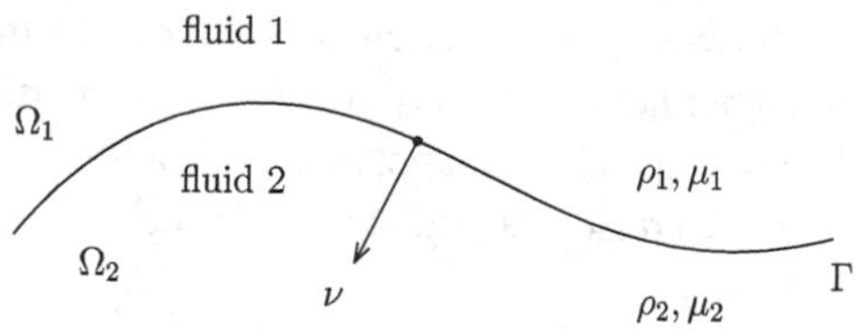

Fig. 3.5 Setting for two phase flow

interface Γ. Either phase has its constant mass density and viscosity, respectively. The model will be derived from basic balance laws. The modeling is rather classical, see for instance [138].

The overall domain Ω is divided into the two phases, $\Omega_1(t)$, $\Omega_2(t)$ and an interface $\Gamma(t)$ separating the two phases. Our model is based on the balance equations for mass and momentum. We proceed analogously as in Sect. 3.6. Consider the mass density of some generic extensive thermodynamical quantity a. Let $U \subset \Omega$ be an open test volume. We assume that in addition to the boundary term, which one gets from Reynolds' transport theorem, there is a diffusive flux q_d at the boundary and in turn the rate of change of ρa in U is given by

$$\frac{d}{dt}\int_U \rho a \, dx = \int_{\partial U} -(\rho a u + q_d)\cdot n \, d\mathcal{H}^{n-1} + \int_U f \, dx + \int_{\Gamma\cap U} f_\Gamma \, d\mathcal{H}^{n-1}. \qquad (3.32)$$

Here, ρ denotes the density of the fluid, u the fluid velocity, q_d denotes the diffusive flux of a, f denotes bulk sources and sinks, f_Γ is a possible boundary density of sources and sinks and n is the outward pointing unit normal of ∂U.

Assuming the open test volume U to be completely contained in one of the phases $\Omega_1(t)$, $\Omega_2(t)$, applying Gauss' theorem to Eq. (3.32) yields

$$\int_U (\partial_t(\rho a) + \nabla\cdot(q_d + \rho a u) - f) \, dx = 0. \qquad (3.33)$$

Since Eq. (3.33) holds for arbitrary open test volumes $U \subset (\Omega_1(t) \cup \Omega_2(t))$ it follows that for all t

$$\partial_t(\rho a) + \nabla\cdot(q_d + \rho a u) = f \quad \text{in} \quad \Omega_1(t) \cup \Omega_2(t). \qquad (3.34)$$

We now set $a = 1$, $q_d = 0$ and $f = 0$ in Eq. (3.34) to obtain the equation for the balance of mass:

$$\partial_t \rho + \nabla\cdot(\rho u) = 0 \quad \text{in} \quad \Omega_1(t) \cup \Omega_2(t) \quad \text{for all} \quad t. \qquad (3.35)$$

Combining the last two equations finally yields the generic balance equation

$$\rho(\partial_t a + u\cdot\nabla a) + \nabla\cdot q_d = f \quad \text{in} \quad \Omega_1(t) \cup \Omega_2(t) \quad \text{for all} \quad t. \qquad (3.36)$$

In a similar way, but working with a test volume intersecting Γ and being transported by the flow field one derives the balance equations for the generic quantity a on the interface Γ as in Sect. 3.6.2:

$$[\rho a(u \cdot \nu - V) + q_d \cdot \nu]_{l_1}^{l_2} = f_\Gamma \quad \text{on} \quad \Gamma(t) \quad \text{for all} \quad t. \tag{3.37}$$

Here $[.]_{l_1}^{l_2}$ denotes a jump of a quantity across the interface and we subtract the value in phase l_1 from the values in phase l_2 and ν is the unit normal on $\Gamma(t)$ pointing into $\Omega_2(t)$. We will always assume that the fluid velocity is continuous, i.e., $[u]_{l_1}^{l_2} = 0$, and the interface is transported with the fluid velocity, i.e., $u \cdot \nu = V$. We hence obtain from (3.37)

$$[q_d \cdot \nu]_{l_1}^{l_2} = f_\Gamma \quad \text{on} \quad \Gamma(t) \quad \text{for all} \quad t. \tag{3.38}$$

3.7.1 Conservation of Mass for Individual Species

The equation for the conservation of mass was already derived for the overall mass density of the fluid ρ, see Eq. (3.35). As we assume incompressibility,

$$\nabla \cdot u = 0, \tag{3.39}$$

it follows that (under appropriate initial and boundary conditions) the density in either phase is constant.

3.7.2 Conservation of Momentum

Inserting $a = u$ in Eq. (3.36) yields the equation for the conservation of momentum. According to Cauchy's Theorem [57, 88, 102], stresses acting on the surface S of a test volume are given by $\mathbf{T}\nu_S$, where $\mathbf{T}$ is the symmetric stress tensor and ν_S the outward pointing normal of S. Therefore, $-\mathbf{T}$ defines the diffusive flux of momentum in Eq. (3.36). The stress tensor $\mathbf{T}$ can be decomposed into a volumetric part and a viscous part:

$$\mathbf{T} = -p\mathbf{I} + \boldsymbol{\tau},$$

where p is called the pressure and $\boldsymbol{\tau}$ the viscous stress tensor. We assume a *Newtonian* fluid, i.e., we assume $\boldsymbol{\tau}$ to be of the form

$$\boldsymbol{\tau} = \mu\Big(\nabla u + (\nabla u)^T - \frac{2}{3}(\nabla \cdot u)\mathbf{I}\Big) + \eta(\nabla \cdot u)\mathbf{I} = 2\mu D(u) + (\eta - \frac{2}{3}\mu)(\nabla \cdot u)\mathbf{I} = 2\mu D(u)$$

where $\mu \geq 0$ is the dynamic viscosity, η is the bulk viscosity and $D(u) = \frac{1}{2}\left(\nabla u + (\nabla u)^T\right)$ is the rate of strain tensor.

Combining all the above leads to the Navier–Stokes equations in either phase:

$$\rho_i(\partial_t u + u \cdot \nabla u) - \mu_i \Delta u + \nabla p = f \qquad \text{in } \Omega_i, \qquad i = 1, 2, \tag{3.40}$$

$$\nabla \cdot u = 0 \qquad \text{in } \Omega_i. \tag{3.41}$$

Here, we have used the fact that for a solenoidal vector field, i.e. $\nabla \cdot u = 0$, it holds: $2\nabla \cdot D(u) = \Delta u$.

3.7.3 Jump Condition at the Interface

Set $a = u$ and $q_d = -\mathbf{T}$. Moreover, we assume, as stated before, the continuity of u across Γ, i.e., $[u]_{l_1}^{l_2} = 0$. Then from Eq. (3.37) we derive the jump condition across Γ:

$$[-\mathbf{T}\nu]_{l_1}^{l_2} = f_\Gamma. \tag{3.42}$$

It remains to determine f_Γ.

3.7.4 Surface Tension

Because of the different bindings, in order to move a molecule from the bulk to the surface, a certain amount of energy is required. This means, if the surface area is enlarged, energy has to be invested. This energy has to be always understood as an energy between two phases (for example between water and air) (Figs. 3.6, 3.7, and 3.8). As a consequence we

Fig. 3.6 Experimental observation leading to an evidence for surface tension. By pulling on the stirrup, work is performed against the surface tension. The surface tension can then be calculated from the pulling force on the stirrup before the liquid film breaks off. The figure is taken from https://commons.wikimedia.org/w/index.php?curid=930591

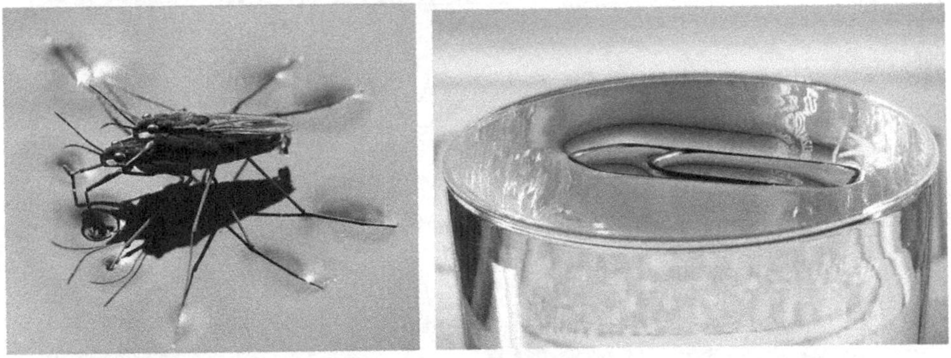

Fig. 3.7 Experimental observations as an evidence for surface tension. The surface tension of water carries water striders and a paper clip. The picture of the water striders was taken by Markus Gayda, see https://de.wikipedia.org/wiki/Datei:Wasserl%C3%A4ufer_bei_der_Paarung_crop.jpg for details. For the paper clip see https://www.wikiwand.com/de/Oberfl%C3%A4chenspannung#Media/Datei:Surface_Tension_01.jpg

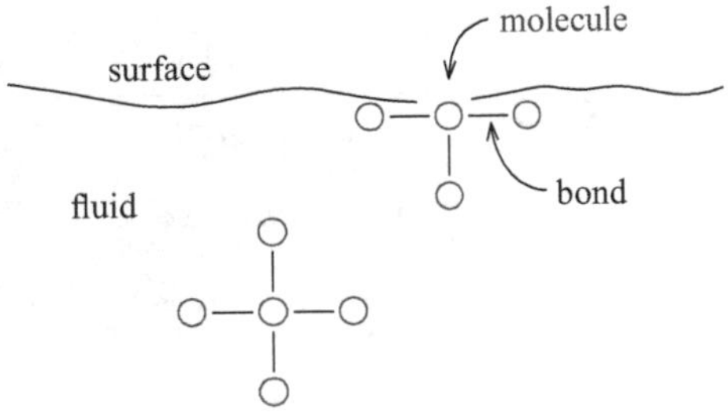

Fig. 3.8 Simple model for an explanation of surface tension

note that

the surface as such has an energy.

In the simplest case we have

Surface energy $\sim$ area of the surface.

So, let $\hat{\Gamma}$ be a possible interface between the two phases. We define the energy

$$E_\gamma(\hat{\Gamma}) = \int_{\hat{\Gamma}} \gamma \, d\mathscr{H}^{n-1}$$

with $\gamma > 0$ the coefficient of surface tension. The simplest case is of course to assume $\gamma \equiv$ constant. We refer to Fig. 3.9 for values of the surface tension coefficient of different materials (for an interface between these materials and air). However, γ may depend on many parameters like temperature or density of a surfactant. For the temperature

Fig. 3.9 Values of the surface tension coefficient γ for some materials at 20 °C

fluids	surface tension in mN/m = 10^{-3}N/m
n-Pentan	16,00
n-Hexan	18,40
Ethanol	22,55
Methanol	22,60
Aceton	23,30
Benzol	28,90
Ethylenglycol	48,4
Glycerin	63,4
water at 80° C	62,6
water at 50° C	67,9
water at 20° C	72,75
mercury at 18° C	471,00
mercury at 20° C	476,00

dependence of the surface tension the famous Eötvös rule says that the surface tension is an affine linear function of the temperature. This rule is approximately fulfilled for most liquids and is especially nearly fulfilled for the water-air interface for temperatures between zero and 100 °C, see Fig. 3.10.

Note that since E_γ has the physical unit of energy, it follows that the physical unit of γ is force by length. We have

$$[E_\gamma] = N \cdot m \qquad \Rightarrow \qquad [\gamma] = \frac{N}{m}.$$

Principle of Virtual Work

How can one derive *forces* f_Γ from energies? We want to use the principle of virtual work to derive the force f_Γ from the energy $E_\gamma(\hat{\Gamma})$. We will first look at a toy example. Consider a mass point x starting at $x = x_0 = 0$ (Fig. 3.11).

Now, x is moved along a curve X from x_0 to x_s against a force field $\tilde{F}$:

$$X : [0, s] \to \mathbb{R}^n, \qquad X(0) = x_0 = 0, \qquad X(s) = x_s.$$

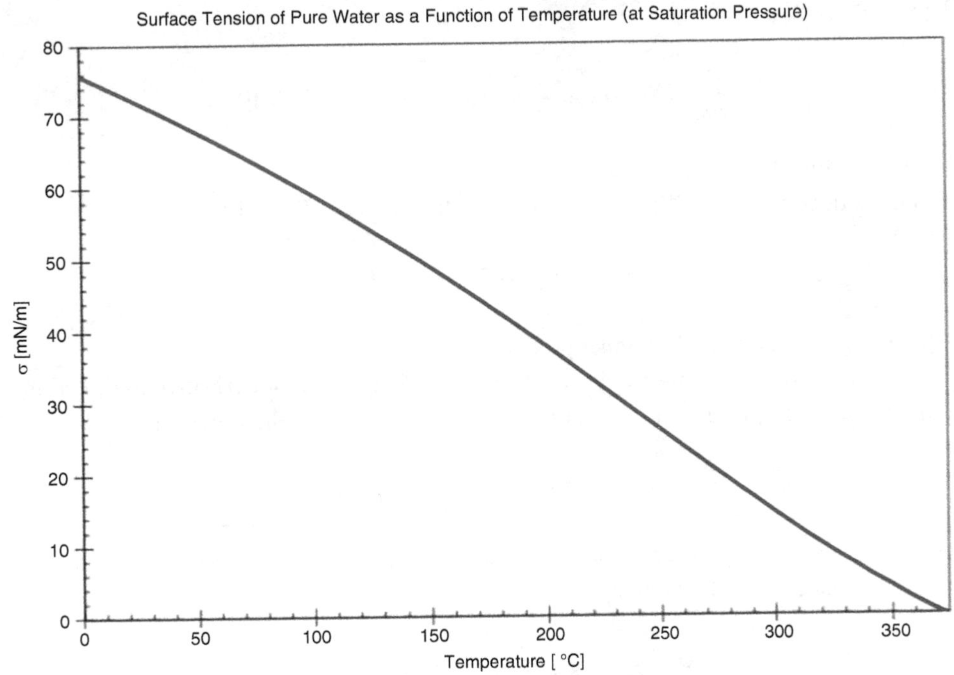

Fig. 3.10 Example: Temperature dependence of the surface tension γ for the water/air interface. The figure is created by Stan J Klimas, see https://commons.wikimedia.org/w/index.php?curid=6742054

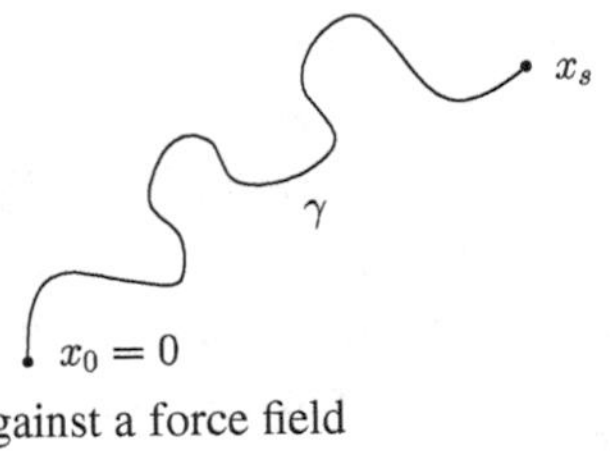

Fig. 3.11 Path of a mass point against a force field

The invested energy is

$$\int_0^s \tilde{F}(X(\tau)) \cdot X'(\tau)\, d\tau = \int_0^s (\tilde{F} \circ X) \cdot \frac{dX}{d\tau}\, d\tau =: \int_0^s (\tilde{F} \circ X) \cdot dX,$$

which means x has gained the energy

$$E(X(s)) - E(x_0) = \int_0^s \tilde{F}(X(\tau)) \cdot X'(\tau)\, d\tau.$$

Upon differentiating we get

$$\frac{d}{ds}E(X(s))_{|s=0} = E'(x_0)X'(0) = \tilde{F}(0) \cdot X'(0).$$

We now define the velocity $v := X'(0)$.

Let us denote the **repelling** force F in $x_0 = 0$ as $F := -\tilde{F}(0)$. Thus

$$E'(x_0)v = -F \cdot v = -\langle F, v \rangle,$$

where $\langle \cdot, \cdot \rangle =$ is the Euclidean inner product.

Now, we look at the other way round: Assume E and its first derivative in direction v ("infinitesimal displacement") for all $v \in \mathbb{R}^n$ are known. Then the force is given by

$$F \cdot v = -E'(x_0)v = -D_v E(x_0). \tag{3.43}$$

We now want to apply the principle of virtual work to derive the surface tension force. As energy we take the surface energy:

$$E_\gamma(\hat{\Gamma}) = \int_{\hat{\Gamma}} \gamma(x)\, d\mathcal{H}^{n-1}(x).$$

The single infinitesimal displacement has to be replaced by the displacement field $\zeta : \hat{\Gamma} \to \mathbb{R}^n$. As in Sect. 3.2 we define perturbed surfaces

$$\Gamma(s) := \{x + s\zeta(x) \mid x \in \hat{\Gamma}\}.$$

When perturbing the surface we also need to transport the surface tension coefficient γ. This we do in the same way as [36] by transporting the values $\gamma(x)$ for $x \in \hat{\Gamma}$ along $x + s\zeta(x)$. This means we obtain

$$D_s\gamma = 0,$$

where D_s is the material time derivative. With

$$\Phi_\zeta(s, x) = x + s\zeta(x) \quad \text{for} \quad x \in \hat{\Gamma}$$

we define

$$\Psi_\zeta(s, \cdot) \quad \text{as the inverse of} \quad \Phi_\zeta(s, \cdot)$$

as well as the perturbed energies

$$E_\gamma(\Gamma(s)) := \int_{\Gamma(s)} \gamma(\Psi_\zeta(s, y))\, d\mathcal{H}^{n-1}(y).$$

As we transport $\gamma(x)$ for $x \in \hat{\Gamma}$ along the path $s \mapsto x + s\zeta(x)$ we obtain

$$\Psi_\zeta(s, y) = x \quad \text{if and only if} \quad y = x + s\zeta(x).$$

This implies for the material time derivative $D_s \Psi_\zeta(s, y) = 0$. On $\hat{\Gamma}$ we now define $V = \zeta \cdot \nu$ and $v_\tau = \zeta - (\zeta \cdot \nu)\nu$ and on $\partial\hat{\Gamma}$ we define the outer unit conormal $\nu_{\partial\hat{\Gamma}}$. We hence obtain, using the fact that the material time derivative vanishes, the Transport Theorem 2.10.1 and the integration by parts formula stated in Proposition 2.5.1

$$\begin{aligned}
\delta E_\gamma(\hat{\Gamma})(\zeta) &= \frac{d}{ds} E_\gamma(\Gamma(s))_{|s=0} \\
&= \int_{\hat{\Gamma}} \gamma(-V\kappa + \nabla_{\hat{\Gamma}} \cdot v_\tau)\, d\mathcal{H}^{n-1} \\
&= -\int_{\hat{\Gamma}} (\gamma V\kappa + \nabla_{\hat{\Gamma}}\gamma \cdot v_\tau)\, d\mathcal{H}^{n-1} + \int_{\partial\hat{\Gamma}} \gamma v_\tau \cdot \nu_{\partial\hat{\Gamma}}\, d\mathcal{H}^{n-2} \\
&= -\int_{\hat{\Gamma}} (\gamma(\zeta \cdot \nu)\kappa + \nabla_{\hat{\Gamma}}\gamma \cdot (\zeta - (\zeta \cdot \nu)\nu))\, d\mathcal{H}^{n-1} \\
&\quad + \int_{\partial\hat{\Gamma}} \gamma(\zeta - (\zeta \cdot \nu)\nu) \cdot \nu_{\partial\hat{\Gamma}}\, d\mathcal{H}^{n-2}.
\end{aligned} \tag{3.44}$$

The expression $-\delta E_\gamma(\hat{\Gamma})(\zeta)$ is the virtual work induced by γ in the direction of ζ. In (3.43) we chose an inner product to define the force.

Question What is the right inner product $\langle \cdot, \cdot \rangle$ in the case of the surface energy E_γ?

Choose

$$\langle F, \zeta \rangle := \int_{\hat{\Gamma}} f_{\hat{\Gamma}} \cdot \zeta\, d\mathcal{H}^{n-1} + \int_{\partial\hat{\Gamma}} f_{\partial\hat{\Gamma}} \cdot \zeta\, d\mathcal{H}^{n-2}.$$

Then from the identities after (3.44) we obtain for the forces $f_{\hat{\Gamma}}$ and $f_{\partial\hat{\Gamma}}$

$$\begin{aligned}
&-\int_{\hat{\Gamma}} (\gamma(\zeta \cdot \nu)\kappa + \nabla_{\hat{\Gamma}}\gamma \cdot (\zeta - (\zeta \cdot \nu)\nu))\, d\mathcal{H}^{n-1} + \int_{\partial\hat{\Gamma}} \gamma(\zeta - (\zeta \cdot \nu)\nu) \cdot \nu_{\partial\hat{\Gamma}}\, d\mathcal{H}^{n-2} \\
&\quad = -\int_{\hat{\Gamma}} f_{\hat{\Gamma}} \cdot \zeta\, d\mathcal{H}^{n-1} - \int_{\partial\hat{\Gamma}} f_{\partial\hat{\Gamma}} \cdot \zeta\, d\mathcal{H}^{n-2}
\end{aligned}$$

for arbitrary smooth $\zeta : \hat{\Gamma} \to \mathbb{R}^n$. Since ζ is arbitrary and $\nu_{\partial\hat{\Gamma}} \in T_{\hat{\Gamma}}$, one concludes

$$f_{\hat{\Gamma}} = \gamma\kappa\nu + \nabla_{\hat{\Gamma}}\gamma \qquad \text{on } \hat{\Gamma}, \tag{3.45}$$

$$f_{\partial\hat{\Gamma}} = -\gamma\nu_{\partial\hat{\Gamma}} \qquad \text{on } \partial\hat{\Gamma}. \tag{3.46}$$

3.7.5 Conditions on the Free Surface Γ

Assume for the moment $\partial\Gamma = \emptyset$, i.e. the free surface is closed. The case with boundary will be treated later. Comparing Eqs. (3.42) and (3.45) one concludes the balance of forces at Γ:

$$[\mathbf{T}\nu]_{l_1}^{l_2} = [2\mu D(u)\nu - p\nu]_{l_1}^{l_2} = -\gamma\kappa\nu - \nabla_\Gamma\gamma. \tag{3.47}$$

This can be written separately for the normal and tangential parts:

$$[\nu \cdot \mathbf{T}\nu]_{l_1}^{l_2} = [2\mu\,(\nu \cdot D(u)\nu) - p]_{l_1}^{l_2} = -\gamma\kappa,$$

$$P[\mathbf{T}\nu]_{l_1}^{l_2} = [2\mu P D(u)\nu]_{l_1}^{l_2} = -\nabla_\Gamma\gamma,$$

where P is the projection onto T_Γ.

Remark If $\gamma = \gamma(\vartheta)$, ϑ the temperature, then by the chain rule

$$\nabla_\Gamma\gamma(\vartheta) = \gamma'(\vartheta)\nabla_\Gamma\vartheta.$$

Thus, a temperature gradient on the interface gives rise to a *tangential* force which in turn induces the so called *Marangoni* flow.

We recall the continuity condition from the beginning:

$$[u]_{l_1}^{l_2} = 0 \qquad \text{on } \Gamma. \tag{3.48}$$

In our problem to compute solutions to an interface problem involving the Navier–Stokes equations we need to take into account that Γ is a *free* boundary, i.e. determining Γ is part of the problem. Therefore, besides (3.47) and (3.48) one further boundary condition is needed. The missing condition is

$$u \cdot \nu = V \qquad \text{on } \Gamma \tag{3.49}$$

with V the normal velocity of Γ. This is called the *kinematic* boundary condition. Note that no condition on the tangential part of u_τ is needed, since u_τ would not change Γ!

3.7.6 The Overall Two-Phase Flow System

We summarize the equations for the two phase flow problem for the case $\partial\Gamma(t) = \emptyset$:

$$\rho_i \cdot (\partial_t u + u \cdot \nabla u) - \mu_i \Delta u + \nabla p = f \qquad \text{in } \Omega_i(t), \qquad i = 1, 2, \tag{3.50}$$

$$\nabla \cdot u = 0 \qquad \text{in } \Omega_i(t) \tag{3.51}$$

with the free boundary conditions on $\Gamma(t)$:

$$[u]_{l_1}^{l_2} = 0, \tag{3.52}$$

$$[2\mu D(u)\nu - p\nu]_{l_1}^{l_2} = -\gamma\kappa\nu - \nabla_\Gamma \gamma, \tag{3.53}$$

$$u \cdot \nu = V. \tag{3.54}$$

The above system has to be completed by initial conditions for u and Γ:

$$u(0, \cdot) = u_0, \qquad \Gamma(0) = \Gamma_0, \tag{3.55}$$

where the last condition is understood in the sense that it also determines the initial shapes of Ω_1 and Ω_2. Moreover, boundary conditions for u on the outer boundary $\partial\Omega$ have to be prescribed. Let us take homogeneous Dirichlet conditions for the latter for simplicity.

Although there is no simple *rigorous* variational formulation with standard energy spaces for this problem, for later purposes it will be beneficial to state the problem in a formal weak form. To this end, like usual, one multiplies the bulk equations by test functions, integrates over Ω, performs an integration by parts and uses the jump conditions on Γ.

The Laplace term and the pressure gradient are handled by:

$$\begin{aligned}
&-\int_{\Omega_1(t)\cup\Omega_2(t)} \mu \Delta u \cdot \varphi \, dx + \int_{\Omega_1(t)\cup\Omega_2(t)} \nabla p \cdot \varphi \, dx \\
&= -\int_{\Omega_1(t)\cup\Omega_2(t)} \nabla \cdot (2\mu D(u)) \cdot \varphi \, dx + \int_{\Omega_1(t)\cup\Omega_2(t)} \nabla p \cdot \varphi \, dx \\
&= \int_{\Omega_1(t)\cup\Omega_2(t)} 2\mu D(u) : D(\varphi) \, dx - \int_{\Omega_1(t)\cup\Omega_2(t)} p \nabla \cdot \varphi \, dx \\
&\quad + \int_{\Gamma(t)} \left([2\mu D(u) - p\mathbf{I}]_{l_1}^{l_2} \nu \right) \cdot \varphi \, d\mathcal{H}^{n-1},
\end{aligned}$$

where the notation

$$\mu = \mu(t,x) = \begin{cases} \mu_1, & (t,x) \in \Omega_1(t), \\ \mu_2, & (t,x) \in \Omega_2(t) \end{cases}$$

and the fact that $\nabla \cdot u = 0$ was used. We also used $2\mu D(u) : (\nabla\varphi) = 2\mu D(u) : D(\varphi)$ which follows from the fact that $D(u)$ is symmetric. Now, using (3.53) and integration by parts one gets

$$\begin{aligned} \int_{\Gamma(t)} [2\mu D(u) - p\mathbf{I}]_{l_1}^{l_2} \nu \cdot \varphi \, d\mathcal{H}^{n-1} &= \int_{\Gamma(t)} \Big(-\gamma\kappa\nu - \nabla_{\Gamma(t)}\gamma \Big) \cdot \varphi \, d\mathcal{H}^{n-1} \\ &= \int_{\Gamma(t)} \gamma \nabla_{\Gamma(t)} \cdot \varphi \, d\mathcal{H}^{n-1}. \end{aligned} \tag{3.56}$$

Thus we arrive at the following formulation.

We define the function spaces $X := (H_0^{1,2}(\Omega))^n$, $Y := L_0^2(\Omega) := \{p \in L^2(\Omega) \mid \int_\Omega p \, dx = 0\}$.

Definition 3.7.1 (Formal Weak Formulation) Find (u, p) with $(u(t), p(t)) \in X \times Y$ for all $t \in [0, T]$ and $(\Gamma(t))_{t\in[0,T]}$ such that $u(0,\cdot) = u_0$, $\Gamma(0) = \Gamma_0$ and fulfilling for all $t \in [0, T]$

$$(\rho(\partial_t u + u \cdot \nabla u), \varphi) + (2\mu D(u), D(\varphi)) - (p, \nabla \cdot \varphi) + \int_{\Gamma(t)} \gamma \nabla_{\Gamma(t)} \cdot \varphi = 0 \qquad \forall \varphi \in X, \tag{3.57}$$

$$(\nabla \cdot u, q) = 0 \qquad \forall q \in Y \tag{3.58}$$

as well as

$$u \cdot \nu = V \qquad \text{on } \Gamma(t). \tag{3.59}$$

Here, $(\cdot,\cdot)$ denotes the L^2 inner product over Ω.

Here, ρ is given by

$$\rho = \rho(t,x) = \begin{cases} \rho_1, & (t,x) \in \Omega_1(t), \\ \rho_2, & (t,x) \in \Omega_2(t), \end{cases}$$

where $\rho_1, \rho_2 > 0$. Note that the continuity of u across Γ is automatically fulfilled, since $u(t) \in X$.

3.7.7 Formal Energy Estimate

We will now derive a (formal) energy estimate fulfilled by weak solutions.

Proposition 3.7.2 (Energy Estimate) *Let $\gamma \equiv const.$, $\partial\Gamma = \emptyset$ and (u, p, Γ) be a sufficiently smooth solution of the formal weak formulation. Then the following energy estimate holds (in differential form):*

$$\frac{1}{2}\frac{d}{dt}||\sqrt{\rho}u(t)||^2 + \gamma\frac{d}{dt}|\Gamma(t)| + 2||\sqrt{\mu}D(u)||^2 = 0,$$

where $||\cdot||$ is the L^2-norm.

Proof

(I) For $i \in \{1, 2\}$ we compute with the help of Reynolds' transport theorem:

$$\frac{1}{2}\frac{d}{dt}\int\limits_{\Omega_i(t)} \rho_i|u(t)|^2\,dx = \frac{1}{2}\int\limits_{\Omega_i(t)} \rho_i\partial_t|u(t)|^2dx \pm \frac{1}{2}\int\limits_{\Gamma(t)} \rho_i|u(t)|^2u\cdot\nu\,d\mathcal{H}^{n-1}$$
$$= \int\limits_{\Omega_i(t)} \rho_i\partial_t u\cdot u\,dx \pm \frac{1}{2}\int\limits_{\Gamma(t)} \rho_i|u(t)|^2u\cdot\nu\,d\mathcal{H}^{n-1},$$

where the sign "$\pm$" depends on i. In fact we have the $+$ for $i = 1$ and $-$ for $i = 2$.

(II) Integration by parts together with incompressibility condition (3.51) gives

$$\int\limits_{\Omega_i(t)} \rho_i(u\cdot\nabla u)\cdot u\,dx = -\int\limits_{\Omega_i(t)} \rho_i(u\cdot\nabla u)\cdot u \pm \int\limits_{\Gamma(t)} \rho_i|u|^2u\cdot\nu\,d\mathcal{H}^{n-1}$$

and therefore we get

$$\pm\frac{1}{2}\int\limits_{\Gamma(t)} \rho_i|u(t)|^2u\cdot\nu d\mathcal{H}^{n-1} = \int\limits_{\Omega_i(t)} \rho_i(u\cdot\nabla u)\cdot u\,dx.$$

(III) From Theorem 2.10.1 with $f \equiv 1$ and Proposition 2.5.1, which states the divergence theorem on manifolds, one infers, using that $\partial\Gamma(t) = \emptyset$, and (3.59)

$$\frac{d}{dt}|\Gamma(t)| = \frac{d}{dt}\int_{\Gamma(t)} 1\, d\mathcal{H}^{n-1} = -\int_{\Gamma(t)} \kappa V\, d\mathcal{H}^{n-1}$$

$$= -\int_{\Gamma(t)} \kappa u \cdot \nu\, d\mathcal{H}^{n-1} = \int_{\Gamma(t)} \nabla_{\Gamma(t)} \cdot u\, d\mathcal{H}^{n-1}.$$

If we now combine (I), (II) and (III) we obtain

$$\frac{d}{dt}\frac{1}{2}\int_\Omega \rho|u|^2 dx + 2\int_\Omega \mu|D(u)|^2 dx + \gamma\,\frac{d}{dt}|\Gamma(t)|$$

$$= \int_\Omega \rho\big(\partial_t u \cdot u + (u\cdot\nabla u)\cdot u\big)\, dx + 2\int_\Omega \mu D(u) : D(u) dx + \int_{\Gamma(t)} \gamma\, \nabla_{\Gamma(t)} \cdot u\, d\mathcal{H}^{n-1}$$

$$= 0,$$

which follows upon testing Eq. (3.57) with $\varphi = u$ and recalling (3.58). □

Remark 3.7.3

(1) In the case where γ is spatially dependent we obtain a similar energy estimate with $\gamma\frac{d}{dt}|\Gamma(t)|$ replaced by $\frac{d}{dt}\int_{\Gamma(t)} \gamma(x)\, d\mathcal{H}^{n-1}$ if we assume that $D_t\gamma = 0$.
(2) In general the surface tension can depend on some species concentration or the temperature. In this case there is a coupling to equations for the concentration and/or the temperature and instead of $D_t\gamma = 0$ a more complex and more realistic equation has to hold. We refer to [12, 21, 22] for a treatment of such situations where also weak formulations and energy estimates are treated. In such situations a Marangoni convection caused by gradients of surface tension will take place. This is due to the fact that a liquid with a higher γ pulls more strongly on the surrounding liquid than one with a lower γ and the presence of a gradient in surface tension γ will hence cause the liquid to flow to regions with a higher surface tension. As discussed above gradients of surface tension can be caused by concentration gradients or by temperature gradients.

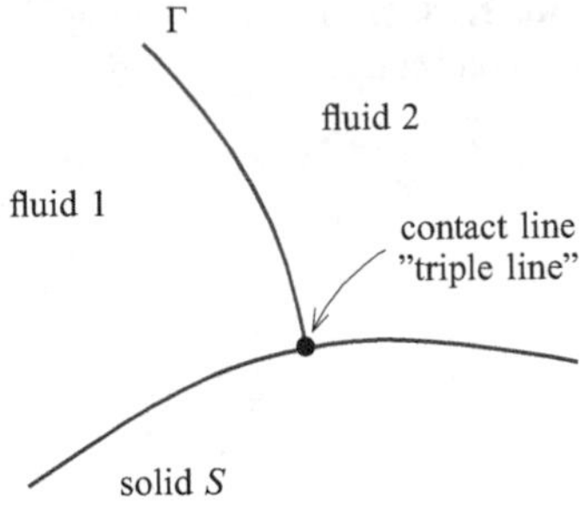

Fig. 3.12 Sketch of the situation and notation for the case with a triple line

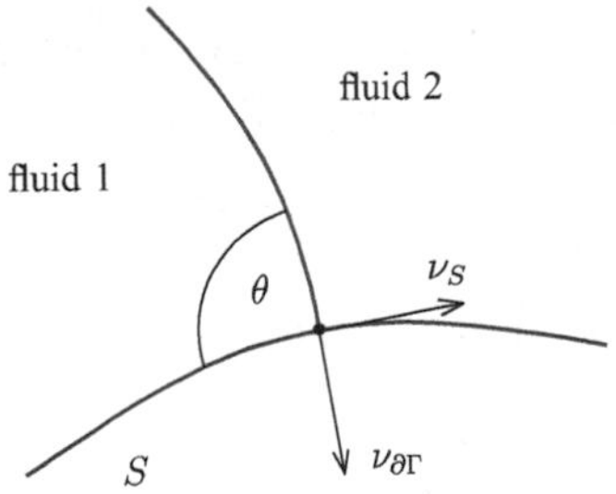

Fig. 3.13 Contact angle

3.7.8 Contact Angle

We now treat the case $\partial\Gamma \neq \emptyset$. The situation is depicted in Figs. 3.12 and 3.13. We make use of condition Eq. (3.46), the force arising in direction $\nu_{\partial\Gamma}$.

As already pointed out, surface tension is always understood in the sense of an interface separating two materials. For the situation in Fig. 3.12 this means:

$$\gamma_{12} = \text{surface tension between fluid 1 and fluid 2,}$$

$$\gamma_{1S} = \text{surface tension between fluid 1 and solid } S,$$

$$\gamma_{2S} = \text{surface tension between fluid 2 and solid } S,$$

and we take $\gamma_{12} = \gamma$.

The contact line or "triple line" is the intersection of all three phases, fluids 1 and 2 as well as S.

We now choose $\nu_S \in T_{\partial S}$, where $T_{\partial S}$ is the tangent space to ∂S, such that $\nu_S \perp$ (contact line).

At the contact line the following forces act (Fig. 3.14): First of all we have a force $f_{\partial\Gamma}$ which by Eq. (3.46) is given as

$$f_{\partial\Gamma} = -\gamma_{12}\nu_{\partial\Gamma}.$$

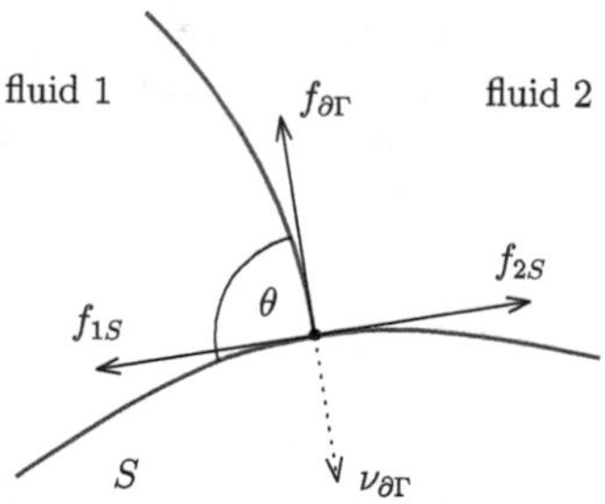

Fig. 3.14 Balance of forces at the contact line

With the same argument that led to Eq. (3.46) we get

$$f_{1S} = -\gamma_{1S}\nu_S \quad \text{on the contact line,}$$

$$f_{2S} = \quad \gamma_{2S}\nu_S \quad \text{on the contact line.}$$

Now we assume that the sum of the forces in tangential directions of the solid vanishes, i.e., the tangential parts of the forces are in equilibrium, so that we have a **balance of forces.** In the normal direction to ∂S no movement resulting from forces is possible and hence the normal part of the forces cannot intrinsically equilibrate. Instead the solid boundary itself will yield a counterforce to guarantee that the whole system is in a force balance.

For the balance of forces it has to hold:

$$(f_{1S} + f_{2S} + f_{\partial\Gamma}) \cdot \nu_S = 0.$$

Therefore:

$$-\gamma_{1S} + \gamma_{2S} - \gamma_{12}\nu_{\partial\Gamma} \cdot \nu_S = 0$$

or

$$-\gamma_{1S} + \gamma_{2S} - \gamma_{12}\cos\theta = 0$$

with θ being the *static contact angle* (which can be measured experimentally).

This leads to the contact angle condition

$$\cos\theta = \frac{\gamma_{2S} - \gamma_{1S}}{\gamma_{12}}.$$

An example is illustrated in Fig. 3.15 where 1 corresponds to water, 2 corresponds to air and S as before stands for the solid.

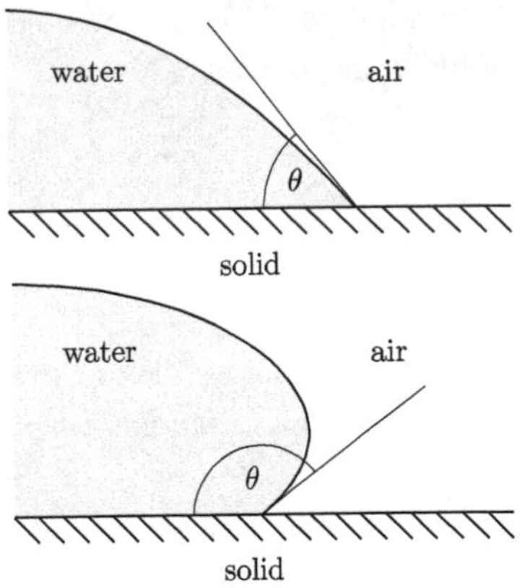

Fig. 3.15 Example of small and large contact angle

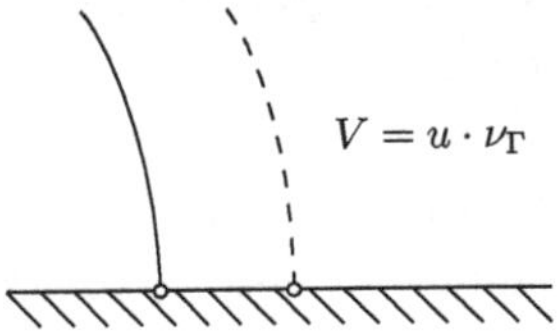

Fig. 3.16 Dirichlet condition leads to a force singularity

We need to impose a boundary condition for u on the wall ∂S. Clearly, since there is no flow through the solid wall, the normal component of u vanishes, i.e.,

$$u \cdot \nu = 0 \text{ on } \partial S,$$

where ν is an outer unit normal on ∂S.

For the tangential part of the velocity, u_τ one might be tempted to also impose a Dirichlet type condition:

$$u_\tau = 0 \qquad \text{on } \partial S.$$

Does this make sense?

In case $\theta \neq 0, \ \theta \neq \pi$ this would imply a jump of u, since

$$V = u \cdot \nu_\Gamma$$

in case the interface moves. Consequently, $u(t, \cdot) \notin H^1$, which is physically not meaningful, since it implies a force singularity (Fig. 3.16).

We remark that finding "correct" boundary conditions close to the contact line is still subject to intensive research.

Some "good" boundary conditions are the Navier boundary conditions close to the contact line which are given as

$$u \cdot \nu = 0 \quad \text{on } \partial S,$$

$$P(2\mu D(u)\nu + \beta u) = 0 \quad \text{on } \partial S$$

with $\beta \geq 0$ and P being the projection onto $T_{\partial S}$. This boundary condition can be formulated in a variational way.

Question What is a good choice of β? — "Answer comes from experiments".

Away from the contact line, one can set β large or even switch to a Dirichlet condition for u.

Remark For a rapidly moving contact line, a static contact angle condition may not be sufficient. Instead, one then should rather impose a *dynamic contact angle*, where the instantaneous contact angle depends on the speed of the contact line, see for instance [39, 95, 98].

3.8 Phase Field Models

In the geometric evolution equations and in the free boundary problems discussed so far, the interface was described as a hypersurface. In the last thirty years, phase field approaches have been another successful approach describing the evolution of interfaces. In particular, phase field methods allow for a change of topology. In a phase field description of interface evolution, instead of a characteristic function $\chi : \Omega \to \{0, 1\}$, which describes the two regions occupied by the phases, one uses a smooth function which takes values close to given values, e.g., ± 1, and rapidly changes between these two values in a small interfacial region.

3.8.1 The Ginzburg–Landau Energy

The phase field approach is best motivated by considering the so-called Ginzburg–Landau energy

$$E_\varepsilon(\varphi) := \int_\Omega \left(\tfrac{\varepsilon}{2}|\nabla\varphi|^2 + \tfrac{1}{\varepsilon}\psi(\varphi)\right)dx, \tag{3.60}$$

Fig. 3.17 The energy contribution $\psi(\varphi)$ in (3.60) penalizes values of φ, which differ from ± 1

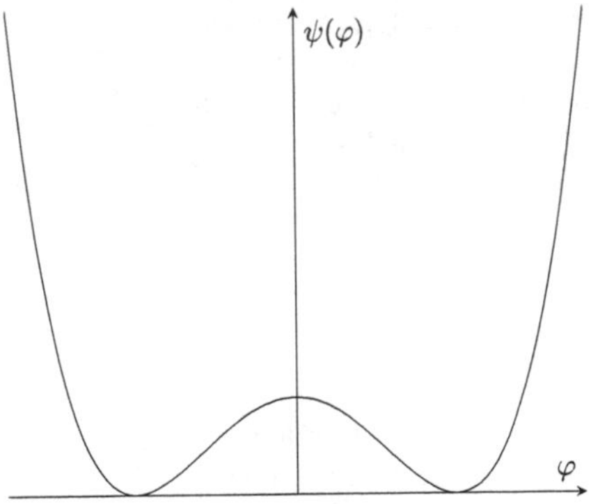

where $\varepsilon > 0$ is a small parameter. For functions φ with a moderate energy $E_\varepsilon(\varphi)$, it will turn out that ε is proportional to the interfacial thickness between the region $\{\varphi \approx -1\}$ and $\{\varphi \approx 1\}$. The function $\psi : \mathbb{R} \to \mathbb{R}_0^+$ is a double-well potential having two global minima with value zero at ± 1, i.e., $\psi(\pm 1) = 0$ and $\psi(z) > 0$ for $z \notin \{-1, 1\}$, see Fig. 3.17 for an example. Typical choices are the quartic potential

$$\psi(\varphi) = \frac{9}{32}(\varphi^2 - 1)^2$$

and the double obstacle potential ψ_{ob}, which is defined as

$$\psi_{ob}(\varphi) = \frac{1}{2}(1 - \varphi^2) \qquad \text{for all} \quad \varphi \in [-1, 1]$$

and ∞ elsewhere, see [27], although different choices are possible. A choice motivated from entropy considerations leads to

$$\psi_{log}(\varphi) = \theta[(1-\varphi)\ln(1-\varphi) + (1+\varphi)\ln(1+\varphi)] + \frac{\theta_c}{2}(1-\varphi^2),$$

see e.g. Abels and Wilke [2]. This function is non-convex with a double-well structure for $\theta < \theta_c$ and the global minima are close to 1 and -1, respectively.

The term $\frac{1}{\varepsilon}\psi(\varphi)$ in the energy E_ε penalizes values which differ from ± 1. In addition, the term $\frac{\varepsilon}{2}|\nabla\varphi|^2$ penalizes gradients of φ and hence too rapid changes of φ in space. It will turn out later that E_ε approximates interfacial energy and that typical solutions of the phase field system have the form illustrated in Fig. 3.18, i.e., they are close to ± 1 in most parts of the domain and have an interfacial region with a thickness which is proportional to ε. In directions normal to the level sets of φ, a typical solution of the phase field system has the form depicted in Fig. 3.19.

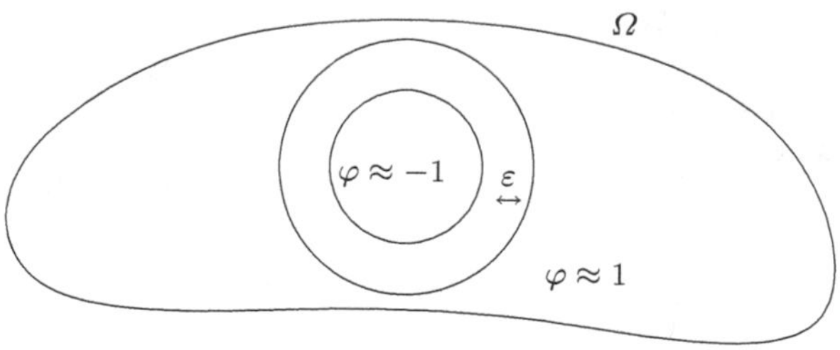

Fig. 3.18 A typical form of the phase field variable φ. Regions in which $\varphi \approx \pm 1$ are separated by a diffuse interfacial layer whose thickness is proportional to ε

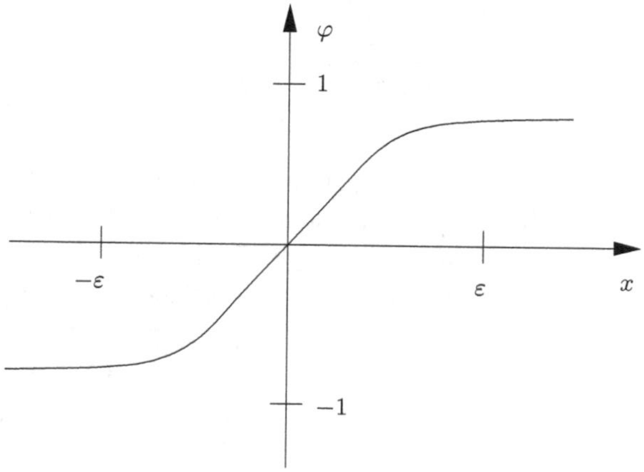

Fig. 3.19 The phase field variable typically has a profile with a phase transition on a diffuse interface of thickness ε

3.8.2 Phase Field Models as Gradient Flows

We now consider different gradient flows involving the energy E_ε. Before discussing the gradient flows, we note that the first variation $\frac{\delta E_\varepsilon}{\delta\varphi}$ of E_ε at $\varphi \in H^1(\Omega)$ in a direction $v \in H^1(\Omega)$ is given by

$$\frac{\delta E_\varepsilon}{\delta \varphi}(\varphi)(v) := \frac{d}{ds} E_\varepsilon(\varphi + sv)_{|s=0} = \int_\Omega (\varepsilon \nabla\varphi \cdot \nabla v + \tfrac{1}{\varepsilon}\psi'(\varphi)\, v)\, dx\,.$$

The Allen–Cahn Equation

Choosing the L^2-inner product for functions defined on Ω, we now obtain the equations for the L^2-gradient flow of E_ε as follows

$$(\partial_t \varphi, v)_{L^2} = -\int_\Omega (\varepsilon \nabla\varphi \cdot \nabla v + \tfrac{1}{\varepsilon}\psi'(\varphi)\, v)\, dx,$$

which has to hold for all times and all suitable test functions v. For functions φ, which are smooth enough, the above is equivalent to

$$\begin{aligned} \partial_t \varphi &= \varepsilon \Delta \varphi - \tfrac{1}{\varepsilon}\psi'(\varphi) && \text{in } (0,T)\times\Omega\,, \\ \frac{\partial \varphi}{\partial n} &= 0 && \text{on } (0,T)\times\partial\Omega, \end{aligned}$$

which follows after integration by parts with the help of the fundamental lemma of the calculus of variations.

The Cahn–Hilliard Equation

It is also possible to consider an H^{-1}-gradient flow of the energy E_ε which preserves the integral of φ. We define

$$H^1_{(m)}(\Omega) = \left\{ u \in H^1(\Omega) \mid ⨍_\Omega u\, dx = m \right\}$$

with $m \in \mathbb{R}$ a given constant. For v_1, v_2 with $\int_\Omega v_i\, dx = 0$, $i = 1, 2$, we define $u_1, u_2 \in H^1_{(0)}(\Omega)$ as weak solutions of

$$\begin{aligned} -\Delta u_i &= v_i \ \text{ in } \Omega\,, \\ \frac{\partial u_i}{\partial n} &= 0 \ \text{ on } \partial\Omega\,. \end{aligned}$$

Note that the above equations have unique solutions in $H^1_{(0)}(\Omega)$. Since the u_i are the solutions of a Neumann problem for the Laplace operator, we set $u_i = (-\Delta_N)^{-1} v_i$. The H^{-1}-inner product is now given as

$$\begin{aligned} (v_1, v_2)_{H^{-1}} &:= \int_\Omega (\nabla(-\Delta_N)^{-1} v_1) \cdot (\nabla(-\Delta_N)^{-1} v_2)\, dx \\ &= \int_\Omega \nabla u_1 \cdot \nabla u_2\, dx = \int_\Omega v_1 u_2\, dx = \int_\Omega v_2 u_1\, dx\,. \end{aligned}$$

For the H^{-1}-gradient flow we have

$$(\partial_t \varphi, v)_{H^{-1}} = -\int_\Omega (\varepsilon \nabla\varphi \cdot \nabla v + \tfrac{1}{\varepsilon}\psi'(\varphi) v)\, dx$$

for test functions $v \in H^1_{(0)}(\Omega)$. Taking the definition of the H^{-1}-inner product into account, we obtain after integration by parts the following boundary value problem:

$$\partial_t \varphi = \Delta(-\varepsilon \Delta \varphi + \tfrac{1}{\varepsilon}\psi'(\varphi)) \qquad \text{in } (0,T)\times\Omega\,, \tag{3.61}$$

$$\tfrac{\partial \varphi}{\partial n} = 0\,, \qquad \tfrac{\partial \Delta \varphi}{\partial n} = 0 \qquad \text{on } (0,T)\times\partial\Omega\,. \tag{3.62}$$

Equation (3.61) is a parabolic partial differential equation of fourth order, which is called the Cahn–Hilliard equation, see [60] and [122] for more details. Solutions of (3.61), (3.62) fulfill

$$\frac{d}{dt}\int_\Omega \varphi\, dx = 0\,, \qquad \frac{d}{dt} E_\varepsilon(\varphi) \le 0\,. \tag{3.63}$$

The fact that the Allen–Cahn equation and the Cahn–Hilliard equation are respectively the L^2- and the H^{-1}-gradient flow of the Ginzburg-Landau energy E_ε, has first been discussed by Fife [69, 70].

The Phase Field System

It is also possible to formulate a phase field analogue of the full Stefan problem (3.25), (3.26), (3.31). We derive a simplified version of the phase field system, similar as in a paper by Penrose and Fife [125] with the help of the gradient flow perspective. To this end, we consider the unknowns internal energy e and phase field φ, for which we define the functional

$$E_\varepsilon(e,\varphi) = \int_\Omega (-s(e,\varphi) + \tfrac{\varepsilon}{2}|\nabla\varphi|^2 + \tfrac{1}{\varepsilon}\psi(\varphi))dx\,,$$

which is related to the negative entropy, see formula (3.14) in [125] for comparison. We now take the inner product $(e_1, e_2)_{H^{-1}} + (\varphi_1, \varphi_2)_{L^2}$ and obtain as gradient flow (not writing down the boundary conditions explicitly)

$$(-\Delta_N)^{-1}\partial_t e = -\frac{\delta E_\varepsilon}{\delta e}\,, \tag{3.64}$$

$$\partial_t \varphi = -\frac{\delta E_\varepsilon}{\delta \varphi}\,. \tag{3.65}$$

Defining $s(e,\varphi) = -\frac{1}{2}(e-\varphi)^2$ and $u = e - \varphi$, we obtain

$$-\frac{\partial s}{\partial e} = u\,, \qquad -\frac{\partial s}{\partial \varphi} = -u$$

and hence we can rewrite (3.64), (3.65) as

$$\partial_t(u+\varphi) = \Delta u\,, \tag{3.66}$$

$$\partial_t\varphi = \varepsilon\Delta\varphi - \tfrac{1}{\varepsilon}\psi'(\varphi) + u\,, \tag{3.67}$$

which has to hold in $(0,T)\times\Omega$. This is the phase field system and u is typically interpreted as temperature or chemical potential.

Parametric Approaches for Geometric Evolution Equations and Interfaces

4

Abstract

In this chapter we will give an overview of different parametric methods for dealing with interfaces. Related to the evolution of interfaces are geometric evolution equations for curves and surfaces. We will present analytical and numerical tools mainly for geometric evolution equations. However, typically they can also be used for more complex models also involving bulk quantities. For parametric models for more complex interface problems we only sketch ideas and do not go into analytical details and refer to the literature for a more precise account of results and proofs.

4.1 Curve Shortening Flow

4.1.1 Local and Global Existence

We consider the evolution of closed curves in $\mathbb{R}^2$ by the curve shortening flow. In the parametric approach to this problem one looks for a mapping $u : [0, T) \times [0, 2\pi] \to \mathbb{R}^2$ such that $x \mapsto u(t, x)$ is 2π-periodic for each $t \in [0, T)$ and

$$u_t = \frac{1}{|u_x|}\Big(\frac{u_x}{|u_x|}\Big)_x \quad \text{in } (0, T) \times [0, 2\pi] \tag{4.1}$$

$$u(0, \cdot) = u_0 \quad \text{in } [0, 2\pi]. \tag{4.2}$$

E. Bänsch et al., *Interfaces: Modeling, Analysis, Numerics*,
Oberwolfach Seminars 51, https://doi.org/10.1007/978-3-031-35550-9_4

Here, $u_0 : [0, 2\pi] \to \mathbb{R}^2$ is a regular parametrization of the given initial curve. A solution of (4.1) clearly satisfies

$$u_t \cdot \nu = \frac{1}{|u_x|}\Big(\frac{u_x}{|u_x|}\Big)_x \cdot \nu = \kappa\nu \cdot \nu = \kappa,$$

which shows that the normal velocity of the curves $\Gamma(t) = u(t, [0, 2\pi])$ is indeed equal to the curvature of $\Gamma(t)$.

Example Let $R > 0$ and $u_0(x) = R\,(\cos(x), \sin(x))$, $x \in [0, 2\pi]$. Then it is easily verified that the function $u(t, x) = \sqrt{R^2 - 2t}\,(\cos(x), \sin(x))$ is a solution of (4.1), (4.2). This shows that a circle shrinks self-similarly to a point in finite time.

More generally, the following result holds:

Theorem 4.1.1 *Let $u_0 : [0, 2\pi] \to \mathbb{R}^2$ be a smooth, embedded closed curve. Then (4.1), (4.2) has a smooth solution on $[0, T)$, which shrinks to a point as $t \nearrow T$. If one rescales the evolving curves in such a way that their enclosed area is constant, then the rescaled curves converge to a circle as $t \nearrow T$.*

Proof For a convex initial curve the result was proved by Gage and Hamilton [75], the generalization to embedded initial curves is due to Grayson [86]. □

Figure 4.1 shows an example of the evolution of a family of convex curves by the curve shortening flow.

The first step in establishing Theorem 4.1.1 consists in proving the existence of a local solution of (4.1), (4.2). Geometric evolution equations like (4.1) typically involve expressions that are invariant with respect to reparametrization which leads to a certain degeneracy in the resulting PDE. In order to make this more precise we write the

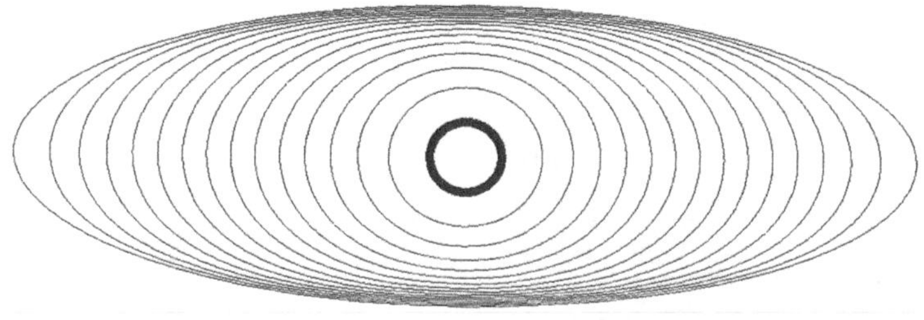

Fig. 4.1 A curve in the plane will shrink to a point in finite time. This is a numerical simulation by Robert Nürnberg, which is taken from [14]

system (4.1) in the form $u_t = A(u_x)u_{xx}$, where $A : \mathbb{R}^2 \setminus \{0\} \to \mathbb{R}^{2\times 2}$ is given by

$$A(p)_{ij} = \frac{1}{|p|^2}\Big(\delta_{ij} - \frac{p_i p_j}{|p|^2}\Big), \quad i, j = 1, 2, \ p \in \mathbb{R}^2 \setminus \{0\}. \tag{4.3}$$

A system of the form $u_t = A(u_x)u_{xx}$ is strictly parabolic provided that the eigenvalues of $A(p)$ have positive real parts for each $p \neq 0$. Clearly, this is not the case for the operator given by (4.3), which has the eigenvalues $\frac{1}{|p|^2}$ and 0 with eigenvectors $p^\perp$ and p respectively. In order to deal with this problem we replace the system (4.1), (4.2) by

$$u_t = \frac{1}{|u_x|}\Big(\frac{u_x}{|u_x|}\Big)_x + \alpha \frac{u_x \cdot u_{xx}}{|u_x|^4} u_x \quad \text{in } (0, T) \times [0, 2\pi], \tag{4.4}$$

$$u(0, \cdot) = u_0 \quad \text{in } [0, 2\pi]. \tag{4.5}$$

Here $\alpha > 0$ is a parameter whose meaning will be explained below. Note that we still have that $u_t \cdot \nu = \kappa$ as the additional term points in tangential direction. We can write (4.4) again in the form $u_t = A_\alpha(u_x)u_{xx}$, where now

$$A_\alpha(p)_{ij} = \frac{1}{|p|^2}\Big(\delta_{ij} + (\alpha - 1)\frac{p_i p_j}{|p|^2}\Big), \quad i, j = 1, 2, \ p \in \mathbb{R}^2, p \neq 0.$$

Since $A_\alpha(p)$ has the eigenvalues $\frac{1}{|p|^2}$ and $\frac{\alpha}{|p|^2}$ the system (4.4) is strictly parabolic and it can be shown that (4.4), (4.5) has a local solution.

In order to understand the relation between (4.4) and the original system (4.1) we assume that u is a solution of (4.4), (4.5) on $[0, T) \times [0, 2\pi]$ and consider the family of ODEs

$$\frac{d}{dt}\eta(t, y) = -\alpha \frac{u_x \cdot u_{xx}}{|u_x|^4}(t, \eta(t, y)), \quad \eta(0, y) = y \tag{4.6}$$

depending on the parameter $y \in \mathbb{R}$. In view of the smoothness of u, the above problem has a unique solution for every $y \in \mathbb{R}$ and the periodicity of u in the spatial variable implies that $\eta(t, y + 2\pi) = \eta(t, y) + 2\pi$ for every $t \in [0, T)$, $y \in \mathbb{R}$. Then, the function $v : [0, T) \times [0, 2\pi] \to \mathbb{R}^2$ with $v(t, y) := u(t, \eta(y, t))$ is 2π-periodic in y and satisfies $v(0, y) = u(0, y) = u_0(y)$. Furthermore

$$v_y(t, y) = \eta_y(t, y)u_x(t, \eta(t, y)), \ v_t(t, y) = \eta_t(t, y)u_x(t, \eta(t, y)) + u_t(t, \eta(t, y)).$$

Combining these relations with (4.4), (4.6) and the fact that the curvature vector $\kappa\nu$ is invariant with respect to reparametrization, we obtain

$$\begin{aligned}
&v_t(t, y) - \frac{1}{|v_y(t, y)|}\Big(\frac{v_y(t, y)}{|v_y(t, y)|}\Big)_y \\
&= \eta_t(t, y)u_x(t, \eta(t, y)) + u_t(t, \eta(t, y)) - \frac{1}{|u_x(t, \eta(t, y))|}\Big(\frac{u_x(t, \eta(t, y))}{|u_x(t, \eta(t, y))|}\Big)_x \\
&= u_t(t, \eta(t, y)) - A_\alpha(u_x(t, \eta(t, y)))u_{xx}(t, \eta(t, y)) \\
&= 0,
\end{aligned}$$

so that v is a solution of (4.1) and $u(t, \cdot)$ and $v(t, \cdot)$ are different parametrizations of the same curve. In order to understand the relation between the two systems a bit better we observe that $|v_y(t, y)| = |\eta_y(t, y)u_x(t, \eta(t, y))| = \eta_y(t, y)|u_x(t, \eta(t, y))|$, where we used that $\eta_y(t, \cdot)$ can be shown to be strictly positive for each $t \in [0, T)$. The chain rule implies

$$\begin{aligned}
&\frac{1}{|v_y(t, y)|}\Big(\frac{\eta_y(t, y)}{|v_y(t, y)|}\Big)_y = \frac{1}{|v_y(t, y)|}\Big(\frac{1}{|u_x(t, \eta(t, y))|}\Big)_y \\
&= -\frac{1}{|v_y(t, y)|}\frac{u_x \cdot u_{xx}}{|u_x|^3}(t, \eta(t, y))\eta_y(t, y) = -\frac{u_x \cdot u_{xx}}{|u_x|^4}(t, \eta(t, y)).
\end{aligned}$$

Recalling (4.6) we find that η is a solution of

$$\eta_t - \alpha\frac{1}{|v_y|}\Big(\frac{\eta_y}{|v_y|}\Big)_y = 0 \quad \text{in } [0, T) \times \mathbb{R}, \tag{4.7}$$

$$\eta(t, y + 2\pi) = \eta(t, y) + 2\pi \quad y \in \mathbb{R}, 0 \le t < T, \tag{4.8}$$

$$\eta(0, y) = y \quad y \in \mathbb{R}. \tag{4.9}$$

Thus, the strictly parabolic system (4.4), (4.5) is equivalent to the system that is obtained by combining (4.1), (4.2) with the parabolic PDE (4.7)–(4.9). This kind of argument was used for the first time for the Ricci flow and is known as the DeTurck trick. We can also use (4.7) in order to gain some insight into the role of the parameter α which occurs as a diffusion parameter in (4.7). Choosing $\alpha = 1$ leads to the particularly simple system

$$u_t = \frac{u_{xx}}{|u_x|^2} \tag{4.10}$$

which we shall use below in order to define and analyze a numerical scheme for the approximation of the curve shortening flow. If we formally consider the limit $\alpha \to \infty$ we expect $\frac{\eta_y}{|v_y|}$ and hence $|u_x|$ to become constant, so that u should be close to a parametrization that is proportional to arclength. We refer to the work of Mikula and Ševčovič [118], Barrett, Garcke and Nürnberg [15, 16, 23] and Elliott and Fritz [61] which use the tangential degrees of freedom and the DeTurck trick in numerical approaches to curvature flow.

4.1.2 Spatial Discretization and Error Analysis

An error analysis for a numerical method based on (4.1) was first carried out by Dziuk in [52]. As mentioned above we use the system (4.10) in order to define our numerical scheme. As we employ a finite element approach we start by writing the system in variational form. To do so, we multiply (4.10) by $|u_x|^2$, then take the scalar product with a test function $\varphi \in H^1_{per}((0, 2\pi); \mathbb{R}^2)$ and integrate by parts. This yields

$$\int_0^{2\pi} u_t \cdot \varphi \, |u_x|^2 dx + \int_0^{2\pi} u_x \cdot \varphi_x dx = 0 \qquad \text{for all } \varphi \in H^1_{per}((0, 2\pi); \mathbb{R}^2). \tag{4.11}$$

In what follows we shall focus on the discretization in space. Let $x_j = jh$, $j = 0, \ldots, J$, where $h = \frac{2\pi}{J}$ denotes the spatial grid size. We approximate our solution/test space $H^1_{per}((0, 2\pi); \mathbb{R}^2)$ by the space of continuous, piecewise linear finite elements, i.e.

$$X_h := \{\varphi_h \in C^0([0, 2\pi]; \mathbb{R}^2) \,|\, \varphi_{h|[x_{j-1}, x_j]} \text{ is affine}, \, j = 1, \ldots, J, \; \varphi_h(0) = \varphi_h(2\pi)\}. \tag{4.12}$$

It is easily verified that $\dim X_h = 2J$, where $\{\varphi_1, \ldots, \varphi_{2J}\}$ with $\varphi_j(x_k) = \delta_{jk} e_1$, $\varphi_{j+J}(x_k) = \delta_{jk} e_2$ $(1 \le j, k \le J)$ is a basis of X_h. In order to specify the approximation properties of X_h we introduce the Lagrange interpolation operator $I_h : C^0_{per}([0, 2\pi]; \mathbb{R}^2) \to X_h$ defined by

$$(I_h f)(x) := \sum_{j=1}^{J} \big(f_1(x_j) \varphi_j(x) + f_2(x_j) \varphi_{j+J}(x) \big), \quad x \in [0, 2\pi].$$

Clearly, $(I_h f)(x_j) = f(x_j)$, $j = 0, \ldots, J$. Furthermore, it is well-known that

$$\| f - I_h f \|_{L^2} + h \| f' - (I_h f)' \|_{L^2} \le c h^2 \| f \|_{H^2} \quad \text{for all } f \in H^2_{per}((0, 2\pi); \mathbb{R}^2). \tag{4.13}$$

Our semi-discrete numerical scheme now reads as follows: find $u_h : [0, T] \times [0, 2\pi] \to \mathbb{R}^2$ such that $u_h(t, \cdot) \in X_h$, $t \in [0, T]$, $u_h(0, \cdot) = I_h u_0$ and

$$\int_0^{2\pi} u_{h,t} \cdot \varphi_h \, |u_{h,x}|^2 dx + \int_0^{2\pi} u_{h,x} \cdot \varphi_{h,x} dx = 0 \qquad \text{for all } \varphi_h \in X_h, 0 < t \le T. \tag{4.14}$$

The following result estimates the error between the solution of (4.10) and (4.14) in terms of the discretization parameter h and was derived in [42].

Theorem 4.1.2 *Suppose that (4.10), (4.5) has a smooth solution on* $[0, T] \times [0, 2\pi]$ *satisfying* $u_x(t, x) \neq 0$, $(t, x) \in [0, T] \times [0, 2\pi]$. *Then there exists* $h_0 > 0$ *such that (4.14) has a unique solution on* $[0, T]$ *and*

$$\max_{t\in[0,T]} \|u_x(t, \cdot) - u_{h,x}(t, \cdot)\|_{L^2} + \left(\int_0^T \|u_t - u_{h,t}\|_{L^2}^2 dt\right)^{\frac{1}{2}} \leq Ch,$$

provided that $0 < h \leq h_0$.

Proof There exist $0 < c_0 < c_1$ such that

$$c_0 \leq |u_x(t, x)| \leq c_1 \quad \text{for all } (t, x) \in [0, T] \times [0, 2\pi]. \tag{4.15}$$

Choose a function $\beta \in C^1([0, \infty))$ such that

$$\beta(s) = s^2, \frac{c_0}{2} \leq s \leq 2c_1, \quad \frac{c_0^2}{4} \leq \beta(s) \leq 4c_1^2, s \geq 0 \text{ and } |\beta'(s)| \leq L, s \geq 0. \tag{4.16}$$

We now consider the following auxiliary problem: find $u_h : [0, T] \times [0, 2\pi] \to \mathbb{R}^2$ such that $u_h(t, \cdot) \in X_h, t \in [0, T], u_h(0, \cdot) = I_h u_0$ and

$$\int_0^{2\pi} u_{h,t} \cdot \varphi_h \, \beta(|u_{h,x}|)dx + \int_0^{2\pi} u_{h,x} \cdot \varphi_{h,x} dx = 0 \qquad \text{for all } \varphi_h \in X_h, 0 < t \leq T. \tag{4.17}$$

By expanding $u_h(t, \cdot) = \sum_{j=1}^{2J} u_j(t)\varphi_j$ we can view (4.17) as a system of ODEs of the form $M(\underline{u})\dot{\underline{u}} + A\underline{u} = 0$, where $\underline{u}(t) = (u_1(t), \ldots, u_{2J}(t))$ and

$$M(\underline{u})_{ij} = \int_0^{2\pi} \varphi_i \cdot \varphi_j \, \beta(|u_{h,x}|)dx, \quad A_{ij} = \int_0^{2\pi} \varphi_{i,x} \cdot \varphi_{j,x} dx, \quad i, j = 1, \ldots, 2J.$$

Using (4.16) it is not difficult to verify that $M(\underline{u})$ is invertible and the system

$$\dot{\underline{u}} = -M(\underline{u})^{-1} A\underline{u}, \quad \underline{u}(0) = (u_{0,1}(x_1), \ldots, u_{0,1}(x_J), u_{0,2}(x_1), \ldots, u_{0,2}(x_J))$$

has a unique local solution on some interval $[0, T_h)$. Choosing $\varphi_h = u_{h,t}(t,\cdot)$ in (4.17) we obtain

$$\int_0^{2\pi} |u_{h,t}|^2 \beta(|u_{h,x}|)dx + \frac{1}{2}\frac{d}{dt}\int_0^{2\pi} |u_{h,x}|^2 dx = 0,$$

from which we infer with the help of (4.16) that

$$\frac{c_0^2}{4}\int_0^t \int_0^{2\pi} |u_{h,t}|^2 dxdt + \frac{1}{2}\int_0^{2\pi} |u_{h,x}(t,\cdot)|^2 dx \\ \leq \frac{1}{2}\int_0^{2\pi} |(I_h u_0)_x|^2 dx \leq c, \quad 0 \leq t < T_h.$$

Using the equivalence of norms on finite dimensional spaces we infer that $\underline{u}$ remains bounded on $[0, T_h)$ so that the solution exists globally in time. Our next goal is to estimate the error between u_h and the exact solution u. Observing that $\beta(|u_x|) = |u_x|^2$ we obtain from taking the difference between (4.11) and (4.17)

$$\int_0^{2\pi} (u_t - u_{h,t}) \cdot \varphi_h \beta(|u_{h,x}|)dx + \int_0^{2\pi} (u_x - u_{h,x}) \cdot \varphi_{h,x} dx \\ = \int_0^{2\pi} u_t \cdot \varphi_h \big(\beta(|u_{h,x}|) - \beta(|u_x|)\big)dx.$$

If we let $\varphi_h = (I_h u_t - u_{h,t})(t,\cdot) \in X_h$ and observe that

$$\int_0^{2\pi} u_x \cdot \varphi_{h,x} dx = \int_0^{2\pi} (I_h u)_x \cdot \varphi_{h,x} dx$$

we deduce with the help of (4.14) and (4.13) that

$$\int_0^{2\pi} |I_h u_t - u_{h,t}|^2 \beta(|u_{h,x}|)dx + \int_0^{2\pi} (I_h u - u_h)_x \cdot (I_h u_t - u_{h,t})_x dx \\ = \int_0^{2\pi} (I_h u_t - u_t) \cdot (I_h u_t - u_{h,t})\beta(|u_{h,x}|)dx \\ + \int_0^{2\pi} u_t \cdot (I_h u_t - u_{h,t})\big(\beta(|u_{h,x}|) - \beta(|u_x|)\big)dx \\ \leq 4c_1^2 \|u_t - I_h u_t\|_{L^2} \|I_h u_t - u_{h,t}\|_{L^2} + L\|u_t\|_{L^\infty} \|I_h u_t - u_{h,t}\|_{L^2} \|u_x - u_{h,x}\|_{L^2} \\ \leq \|I_h u_t - u_{h,t}\|_{L^2} \big(ch^2 \|u_t\|_{H^2} + c\|(I_h u)_x - u_{h,x}\|_{L^2} + ch\|u\|_{H^2}\big) \\ \leq \frac{c_0^2}{8}\|I_h u_t - u_{h,t}\|_{L^2}^2 + ch^2 + c\|(I_h u)_x - u_{h,x}\|_{L^2}^2.$$

Since $\beta(s) \geq \frac{c_0^2}{4}$, $s \geq 0$ we obtain that

$$\frac{c_0^2}{8}\|I_h u_t - u_{h,t}\|_{L^2}^2 + \frac{1}{2}\frac{d}{dt}\|(I_h u)_x - u_{h,x}\|_{L^2}^2 \leq ch^2 + c\|(I_h u)_x - u_{h,x}\|_{L^2}^2 \quad 0 \leq t \leq T. \tag{4.18}$$

Gronwall's lemma implies that

$$\max_{0\leq t\leq T}\|(I_h u - u_h)_x(t,\cdot)\|_{L^2} \leq ch. \tag{4.19}$$

Now, let $I_j = [x_{j-1}, x_j]$ and note that $u_{h,x}$, $(I_h u)_x$ are constant on I_j. Using Taylor approximation together with (4.15) and (4.19) we deduce that

$$\begin{aligned}
|u_{h,x|I_j}| &\geq |(I_h u)_{x|I_j}| - |(I_h u - u_h)_{x|I_j}| \\
&= \left|\frac{u(x_j) - u(x_{j-1})}{h}\right| - \frac{1}{\sqrt{h}}\Big(\int_{I_j}|(I_h u - u_h)_x|^2 dx\Big)^{\frac{1}{2}} \\
&\geq |u_x(x_{j-1})| - ch - \frac{1}{\sqrt{h}}\|(I_h u - u_h)_x\|_{L^2} \geq c_0 - c\sqrt{h} \geq \frac{c_0}{2},
\end{aligned}$$

provided that $0 < h \leq h_0$ and $c\sqrt{h_0} \leq \frac{c_0}{2}$. In a similar way one shows that $|u_{h,x}| \leq 2c_1$ so that $\beta(|u_{h,x}|) = |u_{h,x}|^2$ and u_h solves (4.14). The error bounds now follow from (4.18), (4.19) and (4.13). □

4.1.3 Fully Discrete Scheme and Stability

In order to carry out numerical simulations we also need to discretise in time. To do so, let $t_m = m\delta$, $m = 0, 1, \ldots, M$, where $\delta > 0$ is the time step. Let us denote by u_h^m the approximation of $u(t_m, \cdot)$. Our fully discrete numerical scheme reads as follows: find $u_h^m \in X_h$, $m = 0, 1, \ldots, M$ such that $u_h^0 = I_h u_0$ and

$$\int_0^{2\pi} I_h\left[\frac{u_h^{m+1} - u_h^m}{\delta}\cdot\varphi_h\right]|u_{h,x}^m|^2 dx + \int_0^{2\pi} u_{h,x}^{m+1}\cdot\varphi_{h,x} dx = 0 \qquad \text{for all } \varphi_h \in X_h. \tag{4.20}$$

One can show that (4.20) is equivalent to the following linear system (Exercise 7.10)

$$\frac{1}{2}\big((q_{j+1}^m)^2 + (q_j^m)^2\big)\frac{u_j^{m+1} - u_j^m}{\delta} - \frac{u_{j+1}^{m+1} - 2u_j^{m+1} + u_{j-1}^{m+1}}{h^2} = 0, \quad j = 1, \dots, J. \tag{4.21}$$

Here we have abbreviated

$$u_j^m = u_h^m(x_j) \quad \text{and} \quad q_j^m = \Big|\frac{u_j^m - u_{j-1}^m}{h}\Big|, \quad j = 0, \dots, J,$$

with $u_{J+1}^m = u_1^m$ in view of our periodic setting. The following result shows that the scheme (4.20) decreases both energy and length at the discrete level thus reproducing important properties of the system (4.10).

Lemma 4.1.3 *Suppose that $u_h^m \in X_h$, $m = 0, \dots, M$ is a solution of (4.20). Then*

$$\int_0^{2\pi} |u_{h,x}^{m+1}|^2 dx \le \int_0^{2\pi} |u_{h,x}^m|^2 dx, \; m = 0, \dots, M-1; \tag{4.22}$$

$$\int_0^{2\pi} |u_{h,x}^{m+1}| dx \le \int_0^{2\pi} |u_{h,x}^m| dx, \; m = 0, \dots, M-1. \tag{4.23}$$

Proof If we choose $\varphi_h = u_h^{m+1} - u_h^m$ in (4.20) we obtain

$$\int_0^{2\pi} u_{h,x}^{m+1} \cdot (u_{h,x}^{m+1} - u_{h,x}^m) dx = -\frac{1}{\delta}\int_0^{2\pi} I_h\Big[|u_h^{m+1} - u_h^m|^2\Big] |u_{h,x}^m|^2 dx \le 0.$$

Observing that

$$u_{h,x}^{m+1} \cdot (u_{h,x}^{m+1} - u_{h,x}^m) = \frac{1}{2}|u_{h,x}^{m+1}|^2 - \frac{1}{2}|u_{h,x}^m|^2 + \frac{1}{2}|u_{h,x}^{m+1} - u_{h,x}^m|^2$$

the estimate (4.22) follows. In order to prove (4.23) we write (4.21) in the form

$$\frac{1}{2}\big((q_{j+1}^m)^2 + (q_j^m)^2\big)\frac{u_j^{m+1} - u_j^m}{\delta} = \frac{1}{h}\big(q_{j+1}^{m+1}\tau_{j+1}^{m+1} - q_j^{m+1}\tau_j^{m+1}\big), \tag{4.24}$$

where $\tau_j^{m+1} = (q_j^{m+1})^{-1}\frac{u_j^{m+1}-u_{j-1}^{m+1}}{h}$. Note that $|\tau_j^{m+1}| = 1$. Then we have

$$\int_0^{2\pi} |u_{h,x}^{m+1}|dx - \int_0^{2\pi} |u_{h,x}^{m}|dx = \sum_{j=1}^{J}\big(|u_j^{m+1} - u_{j-1}^{m+1}| - |u_j^m - u_{j-1}^m|\big) \tag{4.25}$$

$$\le \sum_{j=1}^{J} \frac{u_j^{m+1} - u_{j-1}^{m+1}}{|u_j^{m+1} - u_{j-1}^{m+1}|} \cdot \big((u_j^{m+1} - u_{j-1}^{m+1}) - (u_j^m - u_{j-1}^m)\big)$$

$$= \sum_{j=1}^{J} \tau_j^{m+1} \cdot \big((u_j^{m+1} - u_j^m) - (u_{j-1}^{m+1} - u_{j-1}^m)\big),$$

where we used the elementary inequality $|q| \le |p| + \frac{q}{|q|}\cdot(q-p)$ for $p, q \in \mathbb{R}^2$. Summation by parts together with (4.24) gives

$$\sum_{j=1}^{J} \tau_j^{m+1} \cdot \big((u_j^{m+1} - u_j^m) - (u_{j-1}^{m+1} - u_{j-1}^m)\big)$$

$$= -\sum_{j=1}^{J} (\tau_{j+1}^{m+1} - \tau_j^{m+1}) \cdot (u_j^{m+1} - u_j^m)$$

$$= -\sum_{j=1}^{J} \frac{2\delta}{h\big((q_{j+1}^m)^2 + (q_j^m)^2\big)} \big(\tau_{j+1}^{m+1} - \tau_j^{m+1}\big) \cdot \big(q_{j+1}^{m+1}\tau_{j+1}^{m+1} - q_j^{m+1}\tau_j^{m+1}\big)$$

$$= -\sum_{j=1}^{J} \frac{2\delta}{h\big((q_{j+1}^m)^2 + (q_j^m)^2\big)} \big(q_{j+1}^{m+1} + q_j^{m+1}\big)\big(1 - \tau_{j+1}^{m+1} \cdot \tau_j^{m+1}\big)$$

$$= -\sum_{j=1}^{J} \frac{\delta}{h} \frac{q_{j+1}^{m+1} + q_j^{m+1}}{(q_{j+1}^m)^2 + (q_j^m)^2} |\tau_{j+1}^{m+1} - \tau_j^{m+1}|^2 \le 0,$$

where we also used that $1 - p \cdot q = \frac{1}{2}|q - p|^2$ for $|p| = |q| = 1$. If we insert this estimate into (4.25) the bound (4.24) follows. □

4.2 Fully Discrete Anisotropic Curve Shortening Flow

Stability properties that can be proved for a semi-discrete scheme might not hold any more when passing to the full-discretization scheme. This is often the case for highly nonlinear problems where nonlinearities are treated in an explicit way. Instead of choosing very small time steps, it might be advantageous to introduce an artificial "stability term".

We illustrate this idea in the case of the anisotropic curve shortening flow, presenting a scheme that works in any codimension. Because the codimension might be bigger than one we assign a weighting function ϕ (anisotropy map) that acts on the tangent space of the closed curve parametrized by $u : S^1 \to \mathbb{R}^n$, $u = u(x)$, $S^1 \simeq [0, 2\pi]$. In the planar case (where the codimension is equal to one) then one has the correspondence

$$\phi(\tau) = \gamma(\nu)$$

where γ is as in (3.4).

Thus, given a sufficiently smooth norm

$$\phi : \mathbb{R}^n \to [0, \infty)$$

(for more general choices of anisotropy maps, see for instance [128] and comments in there) we consider the anisotropic length functional

$$E_\phi(u) := \int_{S^1} \phi(\tau)|u_x|dx = \int_{S^1} \phi(u_x)dx$$

and the weak formulation of its L^2-gradient flow

$$\int_{S^1} \frac{u_t}{m(\tau)} \cdot \varphi |u_x| dx = - \int_{S^1} \phi'(u_x) \cdot \varphi_x dx \quad \forall \varphi \in H^1(S^1),$$

where $m : S^1 \to (0, \infty)$ is an appropriate mobility factor. For simplicity we choose here $m \equiv 1$. In the isotropic case, where $\phi(p) = |p|$, we recover the classical weak formulation for (4.1). The semi-discrete formulation by piecewise linear finite elements reads

$$\int_{S^1} u_{ht} \cdot \varphi_h |u_{hx}| dx + \int_{S^1} \phi'(u_{hx}) \cdot \varphi_{hx} dx = 0 \qquad \forall \varphi_h \in X_h,$$

where X_h is the space of continuous, piecewise linear finite elements (see (4.12) resp. (4.56)).

A natural full discretization, that treats the nonlinearity in an explicit way is given by

$$\int_{S^1} \frac{u_h^{m+1} - u_h^m}{\delta} \cdot \varphi_h |u_{hx}^m| dx + \int_{S^1} \phi'(\tau_h^m) \cdot \varphi_{hx} dx = 0 \qquad \forall \varphi_h \in X_h \tag{4.26}$$

where δ denotes a given time step, $u_h^m \in X_h$ is the approximation of the parametrization u at the time $t = m\delta$, and $\tau_h^m = \frac{u_{hx}^m}{|u_{hx}^m|}$. Here $\delta M = T$, where $[0, T]$ is the time interval of computation. Moreover we have used the fact that, since the norm ϕ is positively homogeneous of degree one, i.e. $\phi(\lambda p) = \lambda\phi(p)$ for $\lambda > 0$, there holds

$$\phi'(p) \cdot p = \phi(p), \qquad \phi'(\lambda p) = \phi'(p) \qquad \text{for } p \in \mathbb{R}^n \setminus \{0\} \text{ and } \lambda > 0. \tag{4.27}$$

Let us investigate the stability properties of the above scheme, i.e. we are interested in finding out whether the computed length decreases in time.

Testing with $\varphi_h = u_h^{m+1} - u_h^m$, and using the fact that

$$\begin{aligned}
&\phi'(\tau_h^m) \cdot (u_h^{m+1} - u_h^m)_x \\
&= \phi'(u_{hx}^m) \cdot (u_h^{m+1} - u_h^m)_x = \phi'(u_{hx}^m) \cdot u_{hx}^{m+1} - \phi(u_{hx}^m) \qquad \text{by (4.27)} \\
&= \phi(u_{hx}^{m+1}) - \phi(u_{hx}^m) + \phi'(u_{hx}^m) \cdot u_{hx}^{m+1} - \phi(u_{hx}^{m+1}) \\
&= \phi(\tau_h^{m+1})|u_{hx}^{m+1}| - \phi(\tau_h^m)|u_{hx}^m| + |u_{hx}^{m+1}|R^m
\end{aligned}$$

with

$$R^m := \phi'(\tau_h^m) \cdot \tau_h^{m+1} - \phi(\tau_h^{m+1})$$

we infer that

$$\int_{S^1} \frac{|u_h^{m+1} - u_h^m|^2}{\delta} |u_{hx}^m| dx + E_\phi(u_h^{m+1}) - E_\phi(u_h^m) + \int_{S^1} |u_{hx}^{m+1}| R^m dx = 0$$

holds. Thus if $R^m \geq 0$ then $E_\phi(u_h^{m+1}) \leq E_\phi(u_h^m)$ as expected.

However R^m does not have the wished sign. Indeed, even in the isotropic case where $\phi(p) = |p|$ we have

$$R^m = \tau_h^m \cdot \tau_h^{m+1} - 1 = -\frac{1}{2}|\tau_h^m - \tau_h^{m+1}|^2 \leq 0.$$

For a general anisotropy map ϕ one can show (see for example [128, Proposition 7.1]) that

$$R^m \geq -\overline{\gamma}\,|\tau_h^m - \tau_h^{m+1}|^2$$

with $\overline{\gamma} := \frac{1}{\sqrt{5}-1} \max\left\{ \sup_{|p|=1} |\phi'(p)|, \ \sup_{|p|=1} |\phi''(p)| \right\}$.

The idea is now to add an appropriate stability term to (4.26), which can "balance" the negative sign of R^m but does not modify the flow "too much".

Upon noting that

$$|u_{hx}^{m+1} - u_{hx}^m|^2 = |u_{hx}^m||u_{hx}^{m+1}||\tau_h^{m+1} - \tau_h^m|^2 + (|u_{hx}^{m+1}| - |u_{hx}^m|)^2$$

we modify the discrete formulation (4.26) as follows: compute $u_h^{m+1} \in X_h$ such that

$$\int_{S^1} \frac{u_h^{m+1} - u_h^m}{\delta} \cdot \varphi_h |u_{hx}^m| dx + \int_{S^1} \phi'(\tau_h^m) \cdot \varphi_{hx} dx + \sigma \int_{S^1} \phi(\tau_h^m) \frac{u_{hx}^{m+1} - u_{hx}^m}{|u_{hx}^m|} \cdot \varphi_{hx} dx = 0$$

$\forall\, \varphi_h \in X_h$ and for a fixed $\sigma > 0$ such that

$$\sigma \inf_{|p|=1} \phi(p) > \overline{\gamma}.$$

Testing with $\varphi_h = u_h^{m+1} - u_h^m$, repeating the above calculations and summing up over $m \in \{0, \ldots, M\}$ we obtain the wished stability result, that is

$$E_\phi(u_h^0) = \int_{S^1} \phi(\tau_h^0) |u_{hx}^0| dx \geq E_\phi(u_h^M) \quad (+ \text{ other positive terms}).$$

Observe that by writing

$$\sigma \int_{S^1} \phi(\tau_h^m) \frac{u_{hx}^{m+1} - u_{hx}^m}{|u_{hx}^m|} \cdot \varphi_{hx} dx = \delta\sigma \int_{S^1} \frac{\phi(\tau_h^m)}{|u_{hx}^m|} \left(\frac{u_{hx}^{m+1} - u_{hx}^m}{\delta} \right) \cdot \varphi_{hx} dx$$

and interpreting $\frac{u_{hx}^{m+1} - u_{hx}^m}{\delta}$ as an approximation of $u_{tx}(t_m, \cdot)$, we might expect the artificial stability term to have less influence on the flow as $\delta \to 0$.

Finally note that if $\phi(p) = |p|$ (isotropic case) and $\sigma = 1$ then we recover the standard (stable) discretisation of the curve shortening flow

$$\int_{S^1} \frac{u_h^{m+1} - u_h^m}{\delta} \cdot \varphi_h |u_{hx}^m| dx + \int_{S^1} \frac{u_{hx}^{m+1}}{|u_{hx}^m|} \cdot \varphi_{hx} dx = 0 \qquad \forall\, \varphi_h \in X_h.$$

The depicted ideas have been used first in [44] where the approximation of the anisotropic mean curvature flow for hypersurfaces in $\mathbb{R}^n$ in the graph setting has been studied. There, not only stability of the fully discrete scheme, but also its convergence in appropriate norms is shown.

4.3 Mean Curvature Flow

As outlined in Sect. 3.3, a family of hypersurfaces $(\Gamma(t))_{t\in[0,T)}$ in $\mathbb{R}^n$ evolves by mean curvature flow if

$$V = \kappa \quad \text{on } \Gamma(t). \tag{4.28}$$

Here, V and κ are the normal velocity and the mean curvature of $\Gamma(t)$ respectively. Mean curvature flow is a subject that has been studied intensively since the middle of the 80s of the last century and we refer to the books by Ecker [58] and Mantegazza [116] for an overview of the topic.

4.3.1 Some Properties of Solutions

A simple explicit solution can be constructed by looking for hypersurfaces of the form

$$\Gamma(t) = R(t)\mathbb{S}^{n-1},$$

where $\mathbb{S}^{n-1}$ is the unit sphere. Choosing the outer unit normal to the sphere we obtain

$$\kappa = -\frac{n-1}{R(t)}, \qquad V = R'(t),$$

so that the relation $V = \kappa$ leads to the ODE

$$R'(t) = -\frac{n-1}{R(t)}.$$

If we impose the initial condition $R(0) = R_0$ for some $R_0 > 0$ we obtain

$$R(t) = \sqrt{R_0^2 - 2(n-1)t}\,,$$

so that the solution shrinks to a point at time $T = \frac{R_0^2}{2(n-1)}$. The next result shows that we can use the above explicit solution as a comparison hypersurface for a general smooth solution of (4.28).

Theorem 4.3.1 *Let $(\Gamma(t))_{t\in[0,T)}$ be a family of smoothly evolving closed (i.e. compact, without boundary) hypersurfaces that satisfy (4.28) and suppose that $\Gamma(0) \subset \overline{B_{R_0}(0)}$. Then $\Gamma(t) \subset \overline{B_{R(t)}(0)}$, where $R(t) = \sqrt{R_0^2 - 2(n-1)t}$.*

Proof Let us define $\psi : [0,\infty) \to \mathbb{R}$ by

$$\psi(r) := \begin{cases} 0 & , 0 \le r \le R_0^2, \\ (r - R_0^2)^3 & , r > R_0^2. \end{cases}$$

Using Theorem 2.10.1 we infer that

$$\begin{aligned}
&\frac{d}{dt}\int_{\Gamma(t)} \psi(|x|^2+2(n-1)t)d\mathcal{H}^{n-1} \\
&= \int_{\Gamma(t)} \psi'(|x|^2+2(n-1)t)\,\partial_t^{\square}\big(|x|^2+2(n-1)t\big)d\mathcal{H}^{n-1} \\
&\quad - \int_{\Gamma(t)} \psi(|x|^2+2(n-1)t)V\kappa\, d\mathcal{H}^{n-1}.
\end{aligned} \tag{4.29}$$

Since $V\nu = \kappa\nu = \Delta_{\Gamma(t)}x$ by Proposition 2.3.4(ii) we find with the help of Remark 2.7.2(iv) that

$$\begin{aligned}
\partial_t^{\square}\big(|x|^2+2(n-1)t\big) &= \partial_t\big(|x|^2+2(n-1)t\big) + V\nu\cdot\nabla\big(|x|^2+2(n-1)\big) \\
&= 2(n-1)+2V\nu\cdot x = 2(n-1)+2\sum_{i=1}^n x_i\Delta_{\Gamma(t)}x_i \\
&= 2\sum_{i=1}^n\Big[|\nabla_{\Gamma(t)}x_i|^2 + x_i\Delta_{\Gamma(t)}x_i\Big] = \Delta_{\Gamma(t)}|x|^2,
\end{aligned}$$

where we used that $\nabla_{\Gamma(t)}x_i = e_i-(e_i\cdot\nu)\nu$. If we insert this relation into (4.29), integrate by parts and use once more that $V=\kappa$ we deduce that

$$\begin{aligned}
&\frac{d}{dt}\int_{\Gamma(t)} \psi(|x|^2+2(n-1)t)d\mathcal{H}^{n-1} \\
&= \int_{\Gamma(t)} \psi'(|x|^2+2(n-1)t)\,\Delta_{\Gamma(t)}|x|^2 d\mathcal{H}^{n-1} - \int_{\Gamma(t)} \psi(|x|^2+2(n-1)t)\kappa^2\, d\mathcal{H}^{n-1} \\
&= -\int_{\Gamma(t)} \psi''(|x|^2+2(n-1)t)\,\Big|\nabla_{\Gamma(t)}|x|^2\Big|^2 d\mathcal{H}^{n-1} \\
&\quad - \int_{\Gamma(t)} \psi(|x|^2+2(n-1)t)\kappa^2\, d\mathcal{H}^{n-1} \\
&\le 0,
\end{aligned}$$

since $\psi\ge 0$ and $\psi''\ge 0$. This implies that

$$0 \le \int_{\Gamma(t)} \psi(|x|^2+2(n-1)t)d\mathcal{H}^{n-1} \le \int_{\Gamma(0)} \psi(|x|^2)d\mathcal{H}^{n-1} = 0,$$

in view of the fact that $|x|^2\le R_0^2$ for all $x\in\Gamma(0)$. Hence we deduce that $|x|^2+2(n-1)t\le R_0^2$ for all $x\in\Gamma(t)$ which completes the proof. □

Evolution equations for geometric quantities play a crucial role in the analysis of mean curvature flow, see e.g. [92] for the flow of convex hypersurfaces. For later use we derive corresponding equations for the normal ν and the mean curvature κ. If we combine (3.8), (3.14) and the fact that $V = \kappa$ we obtain

$$\partial_t^{\square}\nu = -\nabla_\Gamma V = -\nabla_\Gamma \kappa = \Delta_\Gamma \nu + |\nabla_\Gamma \nu|^2 \nu, \quad \text{on } \Gamma(t), \tag{4.30}$$

while (3.16) yields

$$\partial_t^{\square}\kappa = \Delta_\Gamma V + |\nabla_\Gamma \nu|^2 V = \Delta_\Gamma \kappa + |\nabla_\Gamma \nu|^2 \kappa, \quad \text{on } \Gamma(t). \tag{4.31}$$

The next result shows how to use (4.31) in order to prove that mean convexity is preserved during the evolution by mean curvature.

Lemma 4.3.2 *Let $(\Gamma(t))_{t\in[0,T)}$ be a family of smoothly evolving closed hypersurfaces that satisfy (4.28) and suppose that $\kappa(0,\cdot) \geq 0$ on $\Gamma(0)$. Then $\kappa(t,\cdot) \geq 0$ on $\Gamma(t)$ for $0 \leq t < T$.*

Proof Let $\kappa^-(t,x) := \min(\kappa(t,x), 0)$, $x \in \Gamma(t)$, $0 \leq t < T$. It is not difficult to see that the normal time derivative $\partial_t^{\square}\kappa^-$ exists in the weak sense with

$$\partial_t^{\square}\kappa^- = \begin{cases} \partial_t^{\square}\kappa, & \text{if } \kappa < 0, \\ 0, & \text{if } \kappa \geq 0. \end{cases}$$

Then, using Theorem 2.10.1, suitably extended to weakly differentiable functions, (4.31) and integration by parts we find for $0 \leq t \leq t_0 < T$

$$\begin{aligned}
\frac{d}{dt}\int_{\Gamma(t)} (\kappa^-)^2 d\mathcal{H}^{n-1} &= \int_{\Gamma(t)} \left(2\kappa^-\partial_t^{\square}\kappa - (\kappa^-)^2 V\kappa\right) d\mathcal{H}^{n-1} \\
&= 2\int_{\Gamma(t)} \kappa^-\left(\Delta_\Gamma\kappa + |\nabla_\Gamma \nu|^2\kappa\right) d\mathcal{H}^{n-1} - \int_{\Gamma(t)} (\kappa^-)^2\kappa^2 d\mathcal{H}^{n-1} \\
&\leq -2\int_{\Gamma(t)} |\nabla_\Gamma \kappa^-|^2 d\mathcal{H}^{n-1} + 2\int_{\Gamma(t)} |\nabla_\Gamma \nu|^2 (\kappa^-)^2 d\mathcal{H}^{n-1} \\
&\leq \left(\max_{0\leq \tilde{t}\leq t_0} \max_{\Gamma(\tilde{t})} |\nabla_\Gamma \nu|^2\right) \int_{\Gamma(t)} (\kappa^-)^2 d\mathcal{H}^{n-1}.
\end{aligned}$$

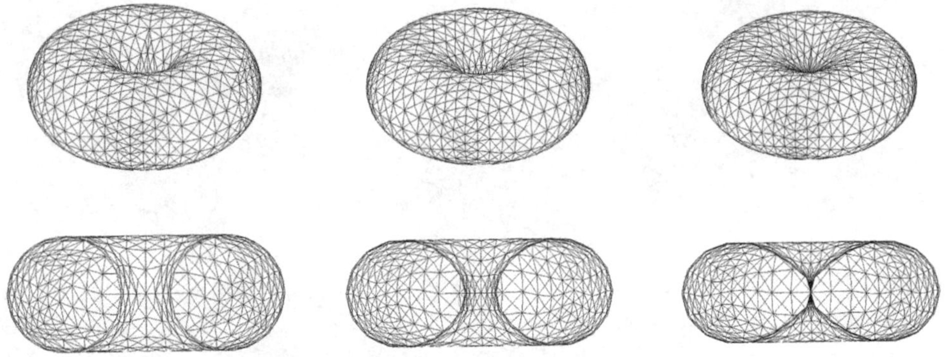

Fig. 4.2 Plots of a solution to mean curvature flow at times $t = 0,\ 0.05,\ 0.09$. A singularity occurs in finite time. This is a numerical simulation by Robert Nürnberg, which is taken from [16]

Gronwall's lemma implies that

$$\int_{\Gamma(t)} (\kappa^-(t,\cdot))^2 d\mathcal{H}^{n-1} \le c \int_{\Gamma(0)} (\kappa^-(0,\cdot))^2 d\mathcal{H}^{n-1} = 0$$

and hence $\kappa^-(t,\cdot) = 0$ on $\Gamma(t), 0 \le t \le t_0$. Since $t_0 < T$ was arbitrary we deduce that $\kappa \ge 0$ on $[0, T)$. □

In the above results properties of solutions were derived under the assumption that the evolution is smooth. In general however, the flow can develop singularities, see Fig. 4.2 for an example.

4.3.2 Existence of Solutions in the Graph Case

We consider a situation where the evolving hypersurface $(\Gamma(t))_{t\in[0,T]}$ can be represented as a graph, i.e.,

$$\Gamma(t) = \{(\hat{x}, h(t,\hat{x}))^T \mid \hat{x} \in U\}$$

where $U \subset \mathbb{R}^{n-1}$ is open and $h : [0, T] \times U \to \mathbb{R}$ is a smooth function.

Lemma 4.3.3 *In the graph case $V = \kappa$ is given by the nonlinear parabolic PDE*

$$\partial_t h = \sqrt{1 + |\nabla_{\hat{x}} h|^2}\, \nabla_{\hat{x}} \cdot \left(\frac{\nabla_{\hat{x}} h}{\sqrt{1 + |\nabla_{\hat{x}} h|^2}} \right). \tag{4.32}$$

Proof Clearly, $\Gamma(t) = \{x \in U \times \mathbb{R} \,|\, \phi(t, x) = 0\}$, where $\phi(t, x) = x_n - h(t, \hat{x})$ and $x = (\hat{x}, x_n)$. Using Proposition 2.3.3 we obtain

$$\nu = \frac{\nabla \phi}{|\nabla \phi|} = \frac{1}{\sqrt{1 + |\nabla_{\hat{x}} h|^2}} (-\nabla_{\hat{x}} h, 1)^T,$$

$$\kappa = -\nabla \cdot \frac{\nabla \phi}{|\nabla \phi|} = \nabla_{\hat{x}} \cdot \left(\frac{\nabla_{\hat{x}} h}{\sqrt{1 + |\nabla_{\hat{x}} h|^2}} \right).$$

As normal velocity we compute

$$V = \partial_t \begin{pmatrix} \hat{x} \\ h(t, \hat{x}) \end{pmatrix} \cdot \nu = \frac{\partial_t h}{\sqrt{1 + |\nabla_{\hat{x}} h|^2}}.$$

Comparing the expressions for V and κ we see that $V = \kappa$ is equivalent to (4.32).

In order to see that (4.32) is parabolic we compute

$$\begin{aligned} \nabla_{\hat{x}} \cdot \left(\frac{\nabla_{\hat{x}} h}{\sqrt{1 + |\nabla_{\hat{x}} h|^2}} \right) &= \sum_{i=1}^{n-1} \partial_i \left(\frac{\partial_i h}{\sqrt{1 + \sum_{j=1}^{n-1} |\partial_j h|^2}} \right) \\ &= \sum_{i,j=1}^{n-1} \left(\frac{\delta_{ij}}{\sqrt{1 + |\nabla_{\hat{x}} h|^2}} - \frac{\partial_i h \partial_j h}{(1 + |\nabla_{\hat{x}} h|^2)^{\frac{3}{2}}} \right) \partial_{ij} h. \end{aligned}$$

Therefore we may write (4.32) in the form

$$\partial_t h = \sum_{i,j=1}^{n-1} a_{ij}(\nabla_{\hat{x}} h) \partial_{ij} h, \qquad \text{where } a_{ij}(p) = \delta_{ij} - \frac{p_i p_j}{1 + |p|^2}. \tag{4.33}$$

It is not difficult to verify that the eigenvalues of the matrix $A(p) = (a_{ij}(p))_{i,j=1}^{n-1}$ are given by 1 and $\frac{1}{1+|p|^2}$, so that $A(p)$ is positive definite for every $p \in \mathbb{R}^{n-1}$. This shows that Eq. (4.32) is strictly parabolic. □

The proof of the above lemma shows that (4.32) is a strictly but not uniformly parabolic PDE. Local existence of solutions is typically shown with the help of Banach's fixed point theorem as can be seen in the following result.

Theorem 4.3.4 *Let $U \subset \mathbb{R}^{n-1}$ be a bounded domain with boundary of class $C^{2+\alpha}$, with $0 < \alpha < 1$, and assume that $h_0 \in C^{2+\alpha}(\bar{U})$ satisfies the compatibility condition*

$$\nabla_{\hat{x}} \cdot \left(\frac{\nabla_{\hat{x}} h_0}{\sqrt{1 + |\nabla_{\hat{x}} h_0|^2}} \right) = 0 \quad \text{on } \partial U.$$

Then, there exists $T > 0$ such that the initial-boundary value problem

$$\begin{aligned} \partial_t h &= \sqrt{1 + |\nabla_{\hat{x}} h|^2}\, \nabla_{\hat{x}} \cdot \left(\frac{\nabla_{\hat{x}} h}{\sqrt{1 + |\nabla_{\hat{x}} h|^2}} \right) && \text{in } (0,T) \times U\,, \\ h(0,x) &= h_0(x) && \text{for } x \in \bar{U}\,, \\ h(t,x) &= h_0(x) && \text{for } t \in (0,T],\, x \in \partial U \end{aligned}$$

has a solution $h \in C^{1+\frac{\alpha}{2}, 2+\alpha}([0,T] \times \bar{U})$.

Here, $C^{1+\frac{\alpha}{2}, 2+\alpha}([0,T] \times \bar{U})$ denotes a parabolic Hölder space. We give a rough sketch of the proof of the above theorem. Let

$$X := \{ h \in C^{1+\frac{\alpha}{2}, 2+\alpha}([0,T] \times \bar{U}) \mid \|h\|_{C^{1+\frac{\alpha}{2}, 2+\alpha}([0,T] \times \bar{U})} \leq M,\ h(x,0) = h_0(x),\ x \in U \}$$

and define the mapping $\mathcal{T} : X \to C^{1+\frac{\alpha}{2}, 2+\alpha}([0,T] \times \bar{U}), \quad g \mapsto h = \mathcal{T}(g)$, where h is the unique solution of the linear parabolic PDE, compare (4.33),

$$\begin{aligned} \partial_t h &= \textstyle\sum_{i,j=1}^{n-1} a_{ij}(\nabla_{\hat{x}} g) \partial_{ij} h && \text{in } (0,T) \times U\,, \\ h(0,x) &= h_0(x) && \text{for } x \in U\,, \\ h(t,x) &= h_0(x) && \text{for } t \in [0,T],\, x \in \partial U. \end{aligned}$$

With the help of linear parabolic theory in Hölder spaces, cf. [106, 110], it can be shown that $\mathcal{T}(X) \subset X$ and that $\mathcal{T}$ is a contraction if T is small and M is chosen sufficiently large. Banach's fixed point theorem then yields a function h with $\mathcal{T}(h) = h$ which is then a solution of our initial-boundary value problem. We refer to [33, 47, 115, 129] for examples where this strategy has been used in similar settings.

4.3.3 Existence in the General Parametric Case

Let us consider the flow of $(n-1)$-dimensional closed (i.e. compact, without boundary) hypersurfaces in $\mathbb{R}^n$. One now looks for a mapping $u : [0, T) \times \Sigma \to \mathbb{R}^n$ such that $u(t, \cdot)$ is a smooth embedding for each $t \in [0, T)$ and

$$\partial_t u = (\kappa\nu) \circ u \quad \text{in } (0, T) \times \Sigma \tag{4.34}$$

$$u(0, \cdot) = u_0 \quad \text{in } \Sigma. \tag{4.35}$$

In the above, Σ is an $(n-1)$-dimensional closed reference hypersurface fixing the genus of the evolving surfaces and ν is a choice of a unit vector of $\Gamma(t) = u(t, \Sigma)$. Furthermore, $u_0 : \Sigma \to \mathbb{R}^n$ denotes a parametrization of the given initial surface Γ_0.

In order to prove the existence of a local solution of (4.34), (4.35) we follow [47,93,116] and describe the hypersurfaces $\Gamma(t)$ as graphs over $\Sigma := \Gamma_0$, i.e. $\Gamma(t) = u(t, \Sigma)$, where

$$u(t, x) = x + \rho(t, x)\nu_\Sigma(x), \quad x \in \Sigma \tag{4.36}$$

and $u_0(x) = x, x \in \Sigma$. Here, ν_Σ denotes the unit normal field to Σ, see Fig. 4.3. It can be shown, see, e.g., [129], that $\Gamma(t)$ is a smooth hypersurface provided that $\rho(t, \cdot)$ is smooth and sufficiently small. Furthermore, the mean curvature and the normal velocity of $\Gamma(t)$ are expressed in terms of ρ via

$$\kappa(t, \cdot) = (\nu(t, \cdot) \cdot \nu_\Sigma)\Delta_{\Gamma(t)}\rho(t, \cdot) + a\big(\cdot, \rho(t, \cdot), \nabla_{\Gamma(t)}\rho(t, \cdot)\big),$$
$$V(t, \cdot) = \partial_t \rho(t, \cdot)(\nu(t, \cdot) \cdot \nu_\Sigma),$$

see the proof of Theorem 1.5.1 in [116]. Using similar ideas as in the graph case one can show that the nonlinear parabolic initial value problem

$$\partial_t \rho - \Delta_{\Gamma(t)}\rho - \frac{1}{\nu \cdot \nu_\Sigma} a(\cdot, \rho, \nabla_{\Gamma(t)}\rho) = 0, \quad \text{in } (0, T] \times \Sigma$$
$$\rho(0, x) = 0, \quad x \in \Sigma$$

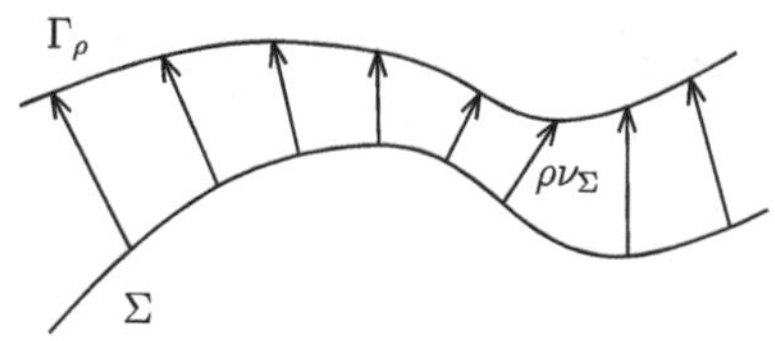

Fig. 4.3 We parametrize $\Gamma_\rho = \{x + \rho(t, x)\nu_\Sigma(x) \mid x \in \Sigma\}$ over Σ

has a unique solution on some small time interval $[0, T]$. The function u given by (4.36) then satisfies

$$\begin{aligned}\partial_t u &= \partial_t \rho\, \nu_\Sigma = \partial_t \rho (\nu \cdot \nu_\Sigma)\nu + \partial_t \rho \big(\nu_\Sigma - (\nu \cdot \nu_\Sigma)\nu\big)\\ &= \big((\nu \cdot \nu_\Sigma)\Delta_{\Gamma(t)}\rho + a(\cdot, \rho, \nabla_{\Gamma(t)}\rho)\big)\nu + \partial_t \rho\big(\nu_\Sigma - (\nu \cdot \nu_\Sigma)\nu\big)\\ &= \kappa\nu + \partial_t \rho\big(\nu_\Sigma - (\nu \cdot \nu_\Sigma)\nu\big).\end{aligned}$$

Since $\partial_t \rho\big(\nu_\Sigma - (\nu \cdot \nu_\Sigma)\nu\big) \cdot \nu = 0$ we see that $V = \partial_t u \cdot \nu = \kappa$ so that the hypersurfaces $(\Gamma(t))_{t\in[0,T]}$ evolve by mean curvature. Finally, by using the same sort of reparametrization argument as described for the curve shortening flow, one obtains a solution of (4.34), (4.35).

4.3.4 Discretization

In this section we restrict ourselves to a brief description of some ideas to solve (4.34), (4.35) numerically. For simplicity we assume in what follows that $n = 3$ and present some key ideas which go back to Dziuk, see [51]. To do so, we define the velocity vector $v(t, \cdot) : \Gamma(t) \to \mathbb{R}^3$ via $v(t, u(t, x)) = u_t(t, x), x \in \Sigma$, so that Proposition 2.3.4 implies that (4.34) can be written in the divergence form

$$v = \kappa\nu = \Delta_{\Gamma(t)}\mathrm{id} \quad \text{on } \Gamma(t). \tag{4.37}$$

We now fix a time step $\delta > 0$ and first describe in an informal way how an approximation of $\Gamma(t + \delta)$ will be obtained from $\Gamma(t)$. Let us abbreviate $u^t := u(t, \cdot)$ and set

$$\hat{u} : \Gamma(t) \to \mathbb{R}^3, \quad \hat{u} := u^{t+\delta} \circ (u^t)^{-1}.$$

Clearly, $\Gamma(t + \delta) = \hat{u}(\Gamma(t))$ and the scheme aims to use this relation at the discrete level. By the definition of v one has

$$v(t, \cdot) \circ u^t = \partial_t u(t, \cdot) \approx \frac{1}{\delta}(u^{t+\delta} - u^t) = \frac{1}{\delta}(\hat{u} - \mathrm{id}) \circ u^t$$

and therefore by (4.37)

$$\frac{1}{\delta}(\hat{u} - \mathrm{id}) \approx v = \Delta_{\Gamma(t)}\mathrm{id} \approx \Delta_{\Gamma(t)}\hat{u} \quad \text{on } \Gamma(t),$$

where the last step is done in order to improve the stability of the resulting scheme. After multiplication by a test function $\varphi \in H^1(\Gamma(t); \mathbb{R}^3)$ and integration by parts on $\Gamma(t)$ the following problem yields an approximation $\tilde{u}$ of $\hat{u} = u^{t+\delta}$:

$$\frac{1}{\delta}\int_{\Gamma(t)} (\tilde{u} - \mathrm{id}) \cdot \varphi \, d\mathcal{H}^2 + \int_{\Gamma(t)} \nabla_{\Gamma(t)}\tilde{u} : \nabla_{\Gamma(t)}\varphi \, d\mathcal{H}^2 = 0 \quad \text{for all } \varphi \in H^1(\Gamma(t); \mathbb{R}^3). \tag{4.38}$$

In order to turn this idea into a numerical method it is necessary to approximate $\Gamma(t)$, which is frequently done with the help of triangular surfaces

$$\Gamma_h = \bigcup_{T \in \mathcal{T}_h} T$$

consisting of space triangles that are either disjoint or intersect in a vertex or an edge. Here, the discretization parameter h refers to a characteristic size (e.g. the diameter) of the triangles belonging to $\mathcal{T}_h$ with the idea that Γ_h converges to $\Gamma(t)$ in a suitable sense as h tends to zero.

In order to formulate Dziuk's algorithm we let $t_m = m\delta, m = 0, 1, 2, \ldots$ and denote by Γ_h^m the triangular approximation of $\Gamma(t_m)$ generated by the scheme.

1. Let Γ_h^0 be a triangular surface approximating Γ_0.
   For m=0,1,2,...
2. Calculate $u_h^{m+1} \in X_h^m$ such that

$$\int_{\Gamma_h^m} \nabla_{\Gamma_h^m} u_h^{m+1} : \nabla_{\Gamma_h^m} \varphi_h \, d\mathcal{H}^2 + \frac{1}{\delta}\int_{\Gamma_h^m} u_h^{m+1} \cdot \varphi_h \, d\mathcal{H}^2$$
$$= \frac{1}{\delta}\int_{\Gamma_h^m} \mathrm{id} \cdot \varphi_h \, d\mathcal{H}^2 \qquad \text{for all } \varphi_h \in X_h^m,$$

   where

$$X_h^m = \{\varphi_h \in C^0(\Gamma_h^m; \mathbb{R}^3) \mid \varphi_{h|T} \text{ is affine on } T \text{ for all } T \in \mathcal{T}_h^m\}.$$

3. Generate the new triangulation $\mathcal{T}_h^{m+1} = \{u_h^{m+1}(T) \mid T \in \mathcal{T}_h^m\}$ and the new surface $\Gamma_h^{m+1} = \bigcup_{T \in \mathcal{T}_h^{m+1}} T$.

Remarks

1. Since u_h^{m+1} is affine on T, it follows that Γ_h^{m+1} is again a triangular surface. The above algorithm is easy to implement as in each time step a linear system of equations needs to be solved. It is possible to extend the method to polynomials of higher order with better approximation properties.

2. It can be shown (see Exercise 7.11) that $|\Gamma_h^{m+1}| \le |\Gamma_h^m|$, reflecting the property of mean curvature flow that it decreases the area during the evolution. Up to now no convergence analysis for the above scheme is available. It is however possible to prove error estimates if one uses polynomials of degree $k \ge 6$, see [108].
3. The above algorithm in some cases leads to mesh distortions during the evolution. We refer to the work of Barrett, Garcke and Nürnberg [15, 16, 23], Mikula and Ševčovič [118] and Elliott and Fritz [61] which discuss numerical approximations to (mean) curvature flow that lead to better meshes.

Let us finish this section by briefly outlining the ideas behind a recently proposed method due to Kovács, Li and Lubich, [104], for which error estimates can be proved. The approach considers an extended system which not only uses the position vector u but also the velocity v, the normal ν and the mean curvature κ as variables. The equation for u is obtained from (4.34) and (4.37) so that

$$\partial_t u(t,p) = v(t,u(t,p)), \quad p \in \Sigma, \quad u(0,\cdot) = u_0 \quad \text{on } \Sigma. \tag{4.39}$$

Next, the equation for the velocity v is incorporated from (4.37) in the form

$$\int_{\Gamma(t)} \left(\nabla_{\Gamma(t)} v : \nabla_{\Gamma(t)} \varphi + v \cdot \varphi\right) d\mathscr{H}^2 = \int_{\Gamma(t)} \left(\nabla_{\Gamma(t)}(\kappa\nu) : \nabla_{\Gamma(t)}\varphi + \kappa\nu \cdot \varphi\right) d\mathscr{H}^2 \tag{4.40}$$

for all $\varphi \in H^1(\Gamma(t), \mathbb{R}^3)$. Finally, one uses the fact that the normal ν and the mean curvature κ of a surface evolving by mean curvature flow satisfy

$$\partial_t^{\square}\nu = -\nabla_\Gamma \kappa = \Delta_\Gamma \nu + |\nabla_\Gamma \nu|^2 \nu, \quad \text{on } \Gamma(t),$$

and

$$\partial_t^{\square}\kappa = \Delta_\Gamma \kappa + |\nabla_\Gamma \nu|^2 \kappa, \quad \text{on } \Gamma(t),$$

compare (4.30) and (4.31). These equations are discretized with the help of the evolving surface finite element method (ESFEM) due to Dziuk and Elliott [53]. Furthermore, a discrete version of (4.40) amounts to an H^1-projection of $\kappa\nu$ to the chosen finite element space. For the resulting semidiscrete scheme using polynomials of order $k \ge 2$ Kovács, Li and Lubich prove that the H^1-error is $O(h^k)$ in all variables (x, v, ν, κ).

4.4 Elastic Flow for Curves

4.4.1 Long Time Existence

To treat fourth order flows new methods must be introduced, since the maximum principle is no longer at our disposal.

Following [55] we give here some ideas on how to show long-time existence for the elastic flow of curves. The procedure we present is based on a combination of L^2-curvature estimates with Gagliardo–Nirenberg-type inequalities and it is a technique which might be adapted to several different situations (see for instance [55, 117], and references given in there).

We parametrize a regular smooth closed curve in $\mathbb{R}^n$ by a periodic map $f : I = \mathbb{R}/\mathbb{Z} \to \mathbb{R}^n$, $f = f(x)$, and we denote by $ds = |\partial_x f| dx$ the arclength element, by $\partial_s = \frac{1}{|\partial_x f|}\partial_x$ the arclength derivative, by $\tau = \partial_s f$ the unit tangent, and by $k = \partial_s^2 f$ the curvature vector. The scalar product in $\mathbb{R}^n$ is denoted by $\langle\, ,\, \rangle$.

The elastic energy [148] is the curvature integral

$$E(f) := \frac{1}{2}\int_I |k|^2 ds := \frac{1}{2}\int_I |k|^2 |\partial_x f| dx \geq 0.$$

(Observe that ds stays for $|\partial_x f| dx$ and we do not actually reparametrize by arc-length. Also we use the notation $\|k\|^2_{L^2(I)} = \int_I |k|^2 ds = \int_I |k|^2 |\partial_x f| dx$.) We call *elasticae* the critical points of E subject to fixed length. Observe that the energy E can be made arbitrarily small, by flattening out a curve at infinity: indeed if we choose f_R to parametrize a circle of radius R, then a quick computation gives $E(f_R) = \frac{\pi}{R} \to 0$ for $R \to \infty$. Therefore it makes sense to consider the modified energy E_λ, which penalizes the length of the curve,

$$E_\lambda(f) := E(f) + \lambda \mathscr{L}(f),$$

where $\lambda > 0$ and $\mathscr{L}(f) = \int_I ds = \int_I |\partial_x f| dx$ is the length functional.

For any variation $f_\varepsilon(x) = f(x) + \varepsilon\phi(x)$, with arbitrary $\phi : I \to \mathbb{R}^n$, one computes

$$\begin{aligned}\frac{d}{d\varepsilon}\Big|_{\varepsilon=0} \mathscr{L}(f_\varepsilon) &= \frac{d}{d\varepsilon}\Big|_{\varepsilon=0} \int_I |\partial_x f + \varepsilon \partial_x \phi| dx = \int_I \langle \tau, \partial_x \phi \rangle dx \\ &= -\int_I \langle \frac{1}{|\partial_x f|}\partial_x \tau, \phi \rangle |\partial_x f| dx = -\int_I \langle k, \phi \rangle ds\end{aligned}$$

and (see Exercise 7.13)

$$\frac{d}{d\varepsilon}\Big|_{\varepsilon=0} E(f_\varepsilon) = \int_I \langle \nabla_s^2 k + \frac{1}{2}|k|^2 k, \phi \rangle ds \tag{4.41}$$

where

$$\nabla_s \phi = \partial_s \phi - \langle \partial_s \phi, \tau \rangle \tau \tag{4.42}$$

denotes the normal component of $\partial_s \phi$. Hence the L^2-gradient flow for E_λ is given by

$$\partial_t f = -\nabla_s^2 k - \frac{1}{2}|k|^2 k + \lambda k. \tag{4.43}$$

By construction

$$\frac{d}{dt} E_\lambda(f) = -\int_I |\partial_t f|^2 ds \tag{4.44}$$

and therefore

$$\frac{1}{2}\|k(t,\cdot)\|^2_{L^2(I)} = E(f(t,\cdot)) \leq E_\lambda(f(t,\cdot)) \leq E_\lambda(f_0). \tag{4.45}$$

Our goal is to sketch some ideas of the following:

Theorem 4.4.1 ([55, Theorem 3.2]) *For any $\lambda \in [0, \infty)$ and smooth initial data f_0, the L^2-gradient flow (4.43) for $E_\lambda(f) = \int_I \left(\frac{1}{2}|k|^2 + \lambda\right) ds$ has a global solution. If $\lambda > 0$, then as $t_i \to \infty$ the curves $f(t_i, \cdot)$ subconverge, when reparametrized by arclength and suitably translated, to an elastica.*

Proof The proof is performed in three steps:

- Step 1 [Short-time existence]:
  Show that given any regular smooth initial data $f_0 : I \to \mathbb{R}^n$, then we can find some $t_0 > 0$ and a smooth $f : [0, t_0) \times I \to \mathbb{R}^n$, $f = f(t, x)$, such that f solves (4.43), $f(0, \cdot) = f_0(\cdot)$, and $f(t, \cdot)$ is regular for all $t \in [0, t_0)$.
- Step 2 [Long-time existence]:
  Let T be the maximal existence time and assume that $0 < T < \infty$. Show that there exist constants c, c_m such that:

$$\sup_{[0,T)} \|\partial_s^m k\|_{L^\infty(I)} \leq c_m = c_m(\Lambda, f_0, T, \lambda) \text{ for all } m \in \mathbb{N}_0, \tag{4.46}$$

$$c^{-1} < |\partial_x f(t,x)| < c \quad \forall\, (t,x) \in [0,T) \times I \text{ with } c = c(\Lambda, f_0, T, \lambda), \tag{4.47}$$

where $\Lambda = 2E_\lambda(f_0)$ denotes the constant that bounds the elastic energy (recall (4.45)), i.e.

$$\sup_{[0,T)} \|k(t,\cdot)\|^2_{L^2(I)} < \Lambda. \tag{4.48}$$

Next show that the above uniform bounds of the length element and of the derivatives of the curvature yield uniform bounds of the derivatives of f in the original parametrization, i.e.

$$\sup_{[0,T)} \|\partial_x^m f\|_{L^\infty(I)} \leq \hat{c}_m(\Lambda, f_0, T, \lambda).$$

One can then extend f smoothly up to $[0,T] \times I$ and by the short-time existence result of Step 1 even beyond T, which contradicts the maximality of T. Hence $T = \infty$.

- Step 3 [Subconvergence result]
  Let $\lambda > 0$, i.e. let the length of the curves stay bounded along the evolution. First of all notice that since $\int_I \tau ds = 0$, the Poincaré's inequality gives

$$c \leq \mathcal{L}(f)\|k\|^2_{L^2(I)} \tag{4.49}$$

(see Exercise 7.14, where by elementary computations one can take $c = 1$; for a sharper constant see [55, 2.18]) so that by (4.48) we infer a uniform bound *from below* for the length of the curves. Together with (4.45) we infer

$$\frac{c}{\Lambda} \leq \mathcal{L}(f(t,\cdot)) \leq \frac{\Lambda}{\lambda} \qquad \text{for any } t \in [0,\infty). \tag{4.50}$$

Moreover the bounds of Step 2 can be now modified to find

$$\sup_{[0,T)} \|\partial_s^m k\|_{L^\infty(I)} \leq c_m(\Lambda, \lambda, f_0).$$

Therefore, given $t_i \to \infty$, we can reparametrize $f(t_i,\cdot)$ over the same interval and obtain convergence of (a subsequence of) the curves $f(t_i,\cdot)$ after a suitable translation.

Moreover, setting

$$u(t) := \|\partial_t f\|^2_{L^2(I)} = \int_I |\partial_t f|^2 ds$$

and recalling (4.44) we observe that

$$\int_0^t u(t')dt' = \int_0^t \int_I |\partial_t f|^2 ds dt' = -(E(f(t)) - E(f_0)) < \infty.$$

Hence, $u \in L^1((0,\infty))$. Since one can show that the derivative has bounded oscillations, i.e. $|\dot{u}(t)| \le c(\lambda, f_0)$, it follows that $u(t) \to 0$ as $t \to \infty$, which means that the limit curve is a critical point for E_λ. □

Next we want to focus on the ideas employed to obtain some of the claims of Step 2. For simplicity let us assume $\lambda > 0$, so that the bound on the length (4.50) holds for any curve in $[0, T)$.

On deriving (4.46): Our goal here is to derive uniform estimates of the derivatives of the curvature vector with respect to arclength.

By embedding theory ($W^{1,1}(I) \subset C^0(\bar{I})$) and (4.50) we see that it is sufficient to show

$$\sup_{[0,T)} \|\partial_s^m k\|_{L^2(I)} \le c_m \qquad \forall\, m \in \mathbb{N}_0. \tag{4.51}$$

On the other hand the "natural operator" appearing in the equation defining the flow (4.43) is not ∂_s but ∇_s. Note however that, for any normal vector field $\phi : I \to \mathbb{R}^n$, i.e. $\langle \phi, \tau \rangle = 0$, we have

$$\nabla_s \phi = \partial_s \phi - \langle \partial_s \phi, \tau \rangle \tau = \partial_s \phi + \langle \phi, k \rangle \tau$$

and therefore $|\partial_s \phi| \le |\nabla_s \phi| + |\langle \phi, k \rangle|$. For instance: $\partial_s k = \nabla_s k - |k|^2 \tau$ and

$$\|\partial_s k\|_{L^2(I)}^2 = \int_I |\partial_s k|^2 ds \le 2\|\nabla_s k\|_{L^2(I)}^2 + 2\int_I |k|^4 ds.$$

Since

$$\int_I |k|^4 ds \le \varepsilon \int_I |\nabla_s k|^2 ds + c\left(\varepsilon, \|k\|_{L^2(I)}, \frac{1}{\mathcal{L}(f)}\right)$$

by an interpolation result [55, Prop. 2.5], we see that in view of (4.50) and (4.48), it is sufficient to have a bound on $\|\nabla_s k\|_{L^2}$ in order to bound the full derivative $\|\partial_s k\|_{L^2}$. This idea extends also to higher derivatives of k, i.e. to obtain (4.51) it suffices to obtain

$$\sup_{[0,T)} \|\nabla_s^m k\|_{L^2(I)} \le c_m \qquad \forall\, m \in \mathbb{N}_0. \tag{4.52}$$

Thus we are left with the task of showing the boundedness of the L^2-norm of the normal vector fields $\nabla_s^m k$. In other words we need to study

$$\begin{aligned}\frac{d}{dt}(\|\phi\|^2_{L^2(I)}) &= 2\int_I \langle \phi, \partial_t \phi\rangle ds + \int_I |\phi|^2 (ds)_t \\ &= 2\int_I \langle \phi, \nabla_t \phi\rangle ds - \int_I |\phi|^2 \langle k, V\rangle ds\end{aligned} \tag{4.53}$$

where $\nabla_t \phi = \partial_t \phi - \langle \partial_t \phi, \tau\rangle \tau$, $V = \partial_t f = -\nabla_s^2 k - \frac{1}{2}|k|^2 k + \lambda k$ is the normal velocity vector, and $\phi = \nabla_s^m k$, for $m \in \mathbb{N}_0$. For $m = 0$ we already know (4.48), therefore there is nothing to prove. Nevertheless, we observe that a direct computation gives (see Exercise 7.12)

$$\begin{aligned}\nabla_t k &= \nabla_s^2 V + \langle k, V\rangle k \\ &= -\nabla_s^4 k - \frac{1}{2}\nabla_s^2(|k|^2 k + \lambda k) + \langle k, V\rangle k\end{aligned}$$

in other words

$$\nabla_t k + \nabla_s^4 k = \text{l. o. t. (with derivatives of } k \text{ of order } \leq 2).$$

This pattern holds for any $\phi := \nabla_s^m k$, $m \in \mathbb{N}_0$, that is [55, Lemma 2.3] :

$$\begin{aligned}\nabla_t(\nabla_s^m k) + \nabla_s^4(\nabla_s^m k) =& \text{l. o. t. (with derivatives of order } \leq m+2) \\ & \text{i.e. collection of terms "of order at most like } \nabla_s^2 \phi\text{".}\end{aligned}$$

This motivates the following strategy: starting from

$$\frac{d}{dt}\left(\frac{1}{2}\int_I |\phi|^2 ds\right) + \int_I |\nabla_s^2 \phi|^2 ds = \int_I \langle (\nabla_t + \nabla_s^4)\phi, \phi\rangle ds - \frac{1}{2}\int_I |\phi|^2 \langle k, V\rangle ds$$

(which follows from (4.53) and integration by parts), one can bound the right-hand side by interpolation inequalities ([55, Prop. 2.5] together with (4.48), (4.50)) and exploiting the positive term $\int_I |\nabla_s^2 \phi| ds$, obtaining

$$\frac{d}{dt}\left(\int_I |\phi|^2 ds\right) + \int_I |\nabla_s^2 \phi|^2 \leq c.$$

Then claim (4.52) follows.

On deriving (4.47): To control the length element $|\partial_x f(t,x)|$ on bounded time intervals, one studies

$$\partial_t(|\partial_x f|) = \langle \tau, \partial_x \partial_t f\rangle = -\langle k, V\rangle |\partial_x f|.$$

Since $\| \langle k, V\rangle \|_{L^\infty(I)} \leq c$ by the previous estimates (4.46), it follows

$$\frac{1}{c} \leq |\partial_x f(t,x)| \leq c \qquad \text{on } [0,T) \times I,$$

where $c = c(\Lambda, f_0, T, \lambda)$.

4.4.2 Stability for the Semi-discrete Problem

Next we would like to discretize in space by piecewise linear finite elements the evolution equation

$$\partial_t f = -\nabla_s^2 k - \frac{1}{2}|k|^2 k + \lambda k \tag{4.54}$$

$$= -\partial_s \left(\partial_s k + \frac{3}{2}|k|^2 \partial_s f \right) + \lambda k. \tag{4.55}$$

Many different algorithms are obviously possible: for instance in [55] a mixed scheme based on the formulation in divergence form (4.55) is proposed.

A central questions is: which discretization yields stability? This is important because, as we know, stability is (in a suitable sense) a precursor to convergence.

Fundamental Idea: *derive the first variation in such a way that all operations performed (in particular integration by parts) are admissible in the FE-space of your choice.* Indeed, if we do not follow this principle, it is possible to introduce errors that we might not be able to control.

Since we want to use *piecewise linear FE* to discretize the fourth order flow (4.54), we introduce two variables, $f_h, k_h \in X_h$, one for the parametrization and one for the curvature vector. Here

$$X_h := \left\{ \eta_h \in C^0([0, 2\pi], \mathbb{R}^n) \, : \, \eta_h|_{I_j} \text{ is affine }, \; j = 1, \ldots, N, \; \eta_h(0) = \eta_h(2\pi) \right\} \tag{4.56}$$

where $0 = u_0 < u_1 < \ldots < u_N = 2\pi$ is a partition of $[0, 2\pi] = I$ into subintervals $I_j = [u_{j-1}, u_j]$. As usual we denote by I_h the Lagrange interpolation operator.

To derive a formulation for the semi-discrete scheme we need to rewrite (4.54) as a system of second order problems for f and k. To this end, let us start again from the energy functional

$$E_\lambda(f) = \frac{1}{2}\int_I |k|^2 |f_x| dx + \lambda \int_I |f_x| dx.$$

For variations of type $f_\varepsilon = f + \varepsilon\phi$, with $\phi : I \to \mathbb{R}^n$, we have

$$\frac{d}{d\varepsilon}\Big|_{\varepsilon=0} E_\lambda(f_\varepsilon) = \int_I \langle k, \frac{d}{d\varepsilon}\Big|_{\varepsilon=0} k_\varepsilon\rangle |f_x| dx + \frac{1}{2}\int_I |k|^2 \left(\frac{d}{d\varepsilon}\Big|_{\varepsilon=0} |(f_\varepsilon)_x|\right) dx$$
$$+ \lambda \int_I \langle \frac{f_x}{|f_x|}, \phi_x\rangle dx.$$

On the other hand we know that (for any test function $\psi : I \to \mathbb{R}^n$) a weak formulation for the curvature vector is given by

$$\int_I \langle k, \psi\rangle |f_x| dx = \int_I \langle \left(\frac{f_x}{|f_x|}\right)_x, \psi\rangle dx = -\int_I \langle \frac{f_x}{|f_x|}, \psi_x\rangle dx$$

and therefore (for the perturbed curve f_ε)

$$\int_I \langle k_\varepsilon, \psi\rangle |(f_\varepsilon)_x| dx + \int_I \langle \frac{(f_\varepsilon)_x}{|(f_\varepsilon)_x|}, \psi_x\rangle dx = 0$$

which yields:

$$\int_I \langle \frac{d}{d\varepsilon}\Big|_{\varepsilon=0} k_\varepsilon, \psi\rangle |f_x| dx + \int_I \langle k, \psi\rangle \langle \frac{f_x}{|f_x|}, \phi_x\rangle dx + \int_I \frac{1}{|f_x|}\langle P\phi_x, \psi_x\rangle = 0$$

with $P = I_n - \tau \otimes \tau = I_n - \partial_s f \otimes \partial_s f$ the normal projection.

Thus, by choosing $\psi = k$, we obtain

$$\frac{d}{d\varepsilon}\Big|_{\varepsilon=0} E_\lambda(f_\varepsilon) = -\int_I |k|^2 \langle \frac{f_x}{|f_x|}, \phi_x\rangle dx - \int_I \frac{1}{|f_x|}\langle P\phi_x, k_x\rangle dx$$
$$+ \frac{1}{2}\int_I |k|^2 \langle \frac{f_x}{|f_x|}, \phi_x\rangle dx + \lambda \int_I \langle \frac{f_x}{|f_x|}, \phi_x\rangle dx$$

and a natural weak form of the gradient flow is given by the system:

$$0 = \int_I \langle f_t, \phi\rangle |f_x| dx - \int_I \frac{\langle Pk_x, \phi_x\rangle}{|f_x|} dx - \frac{1}{2}\int_I |k|^2 \langle \frac{f_x}{|f_x|}, \phi_x\rangle dx + \lambda \int_I \langle \frac{f_x}{|f_x|}, \phi_x\rangle dx, \tag{4.57}$$

$$0 = \int_I \langle k, \psi\rangle |f_x| dx + \int_I \langle \frac{f_x}{|f_x|}, \psi_x\rangle dx. \tag{4.58}$$

For later use, we compute $(4.58)_t$ (i.e. the derivative with respect to time of equation (4.58))

$$0 = \int_I \langle k_t, \psi\rangle |f_x| dx + \int_I \langle k, \psi\rangle \langle \frac{f_x}{|f_x|}, f_{xt}\rangle + \langle \frac{P f_{xt}}{|f_x|}, \psi_x\rangle dx. \qquad (4.58)_t$$

Note that by construction

$$(4.57) \Longleftrightarrow \int_I \langle f_t, \phi\rangle |f_x| dx + \langle E'_\lambda(f), \phi\rangle = 0,$$

which yields (with $\phi = f_t$) the energy decay

$$\frac{d}{dt}(E_\lambda(f)) = \langle E'_\lambda(f), f_t\rangle = -\int_I |f_t|^2 |f_x| dx. \qquad (4.59)$$

On the other hand, if we start with the weak formulation (4.57), (4.58), then (4.59) can be retrieved by choosing $\phi = f_t$ in (4.57) and $\psi = k$ in $(4.58)_t$.

Since (unlike in the derivation of (4.54)) no integration by parts was used at any time in deriving (4.57) and (4.58), then the above operation are admissible also for maps f and k, that are merely in H^1 (and not necessarily smooth).

This is why one defines the semi-discrete scheme as follows: find $f_h, k_h : [0, T] \times I \to \mathbb{R}^n$ such that $f_h(t, \cdot), k_h(t, \cdot) \in X_h$, for $0 \le t \le T$, $f_h(0, \cdot) = I_h f(0, \cdot)$ and

$$0 = \int_I I_h(\langle f_{ht}, \phi_h\rangle) |f_{hx}| dx - \int_I \frac{\langle P_h k_{hx}, \phi_{hx}\rangle}{|f_{hx}|} dx - \frac{1}{2}\int_I I_h\left[|k_h|^2\right] \langle \tau_h, \phi_{hx}\rangle dx$$
$$+ \lambda \int_I \langle \tau_h, \phi_{hx}\rangle dx, \qquad (4.60)$$

$$0 = \int_I I_h(\langle k_h, \psi_h\rangle) |f_{hx}| dx + \int_I \langle \tau_h, \psi_{hx}\rangle dx \qquad (4.61)$$

hold for test functions $\phi_h, \psi_h \in X_h$. One can verify (Exercise 7.15) that by choosing $\phi_h = f_{ht}$ into (4.60) and $\psi_h = k_h$ in $(4.61)_t$, i.e. (4.61) differentiated with respect to time, we immediately obtain the following energy decrease

$$\frac{d}{dt}\left\{\frac{1}{2}\int_I I_h\left(|k_h|^2\right) |f_{hx}| dx + \lambda \int_I |f_{hx}| dx\right\} = -\int_I I_h\left(|f_{ht}|^2\right) |f_{hx}| dx. \qquad (4.62)$$

Hence, *no extra work* is needed to derive the above stability result.

The error analysis of the semi-discrete scheme is very technical and it can be found in [45]. For the different algorithm proposed in [55] no analysis (stability or error analysis) is available.

4.5 A General Strategy to Solve Interface Problems Involving Bulk Quantities in a Parametric Setting

The basic idea to solve problems like the Stefan problem or the two-phase flow problem is to use a transformation to a domain with a fixed interface Σ, where $\Gamma(t)$ is parametrized over Σ by means of a height function ρ. This strategy has been discussed already in Sect. 4.3.3 for mean curvature flow. In order to also transform bulk quantities we need the *Hanzawa transform* which we will discuss now following the book of Prüss and Simonett [129].

Assume $\Omega \subset \mathbb{R}^n$ is a bounded domain with a boundary $\partial\Omega$ of class C^2. In addition let $\Gamma \subset \Omega$ be a hypersurface of class C^2 which is assumed to be a boundary of a domain $\Omega_1 \subset\subset \Omega$. We set $\Omega_2 = \Omega \setminus \overline{\Omega_1}$. It can be shown that such a Γ can be approximated by a smooth, i.e., a C^∞ or even an analytic hypersurface Σ, see Section 3.4 of Prüss and Simonett [129] for the precise approximation properties. The approximation can be chosen such that Σ bounds a domain Ω_1^Σ with $\overline{\Omega_1^\Sigma} \subset \Omega$ and we set $\Omega_2^\Sigma = \Omega \setminus \overline{\Omega_1^\Sigma}$.

The hypersurface Σ admits a tubular neighborhood in the following sense. There exists a $\delta > 0$ such that the map

$$\begin{aligned} \Lambda : \quad & \Sigma \times (-\delta, \delta) \to \mathbb{R}^n, \\ \Lambda(p, \eta) &:= p + \eta \nu_\Sigma(p) \end{aligned}$$

is a diffeomorphism from $\Sigma \times (-\delta, \delta)$ onto $\mathrm{im}(\Lambda)$, the image of Λ. The inverse

$$\Lambda^{-1} : \mathrm{im}(\Lambda) \to \Sigma \times (-\delta, \delta)$$

has the components

$$\Lambda^{-1}(x) = (\pi_\Sigma(x), d_\Sigma(x)), \quad x \in \mathrm{im}(\Lambda)\,.$$

Here $\pi_\Sigma(x)$ is the projection of x onto Σ and $d_\Sigma(x)$ is the signed distance from x to Σ, i.e., $|d_\Sigma(x)| = \mathrm{dist}(x, \Sigma)$ and $d_\Sigma(x) < 0$ if $x \in \Omega_1^\Sigma$, $d_\Sigma(x) > 0$ if $x \in \Omega_2^\Sigma$. The size of δ is limited by the curvature of Σ. Similar as in Sect. 4.3.3 we can parametrize the free boundary $\Gamma(t)$ over Σ by means of a height function $\rho(t)$ if $\Gamma(t)$ and Σ are close enough. We obtain

$$\Gamma(t) = \{p + \rho(t, p)\nu_\Sigma(p) \mid p \in \Sigma\}, \quad t \geq 0$$

at least for $t \geq 0$ small and $\Gamma(0)$ close to Σ.

Defining $a = \delta/3$ we extend the diffeomorphism $\rho(t, .)$ to all of $\overline{\Omega}$ by means of

$$\begin{aligned} F_\rho(t,x) &= x + \chi(d_\Sigma(x)/a)\rho(t, \pi_\Sigma(x))\nu_\Sigma(\pi_\Sigma(x)) \\ &=: x + \xi_h(t,x)\,. \end{aligned}$$

Here χ is a smooth cut-off function with $0 \le \chi \le 1$, $\chi(r) = 1$ for $|r| < 1$ and $\chi(r) = 0$ for $|r| > 2$. As

$$F_\rho(t,x) = x \quad \text{for} \quad |d_\Sigma(x)| > 2a$$

we notice that F_ρ only has an effect close to the interface.

The strategy is now to transform the problem from $(\Omega_1(t), \Gamma(t), \Omega_2(t))$ to $(\Omega_1^\Sigma, \Sigma, \Omega_2^\Sigma)$. In doing so the sets on which the differential operators are defined become fixed. However, the differential operators now have a highly nonlinear dependence on the unknown geometry, i.e. on ρ. For example in a transformed Laplace operator the height function ρ enters in a highly nonlinear fashion.

We sketch the strategy for the Mullins–Sekerka-problem which is a simplified version of the Stefan problem

$$-\Delta u = 0 \quad \text{in } \Omega_-(t) \cup \Omega_+(t)\,, \tag{4.63}$$

$$V = -[\nabla u]_-^+ \cdot \nu \quad \text{on } \Gamma(t)\,, \tag{4.64}$$

$$u = \kappa \quad \text{on } \Gamma(t)\,, \tag{4.65}$$

$$\nabla u \cdot n = 0 \quad \text{on } \partial\Omega\,. \tag{4.66}$$

For a given ρ_0 small enough we obtain from

$$V = -[\nabla u]_-^+ \cdot \nu \text{ on } \Gamma(t)$$

the following equation

$$\partial_t \rho + B(\rho)v(\rho) = 0, \;\; \rho(0) = \rho_0 \tag{4.67}$$

where $B(\rho)$ and $v(\rho)$ are defined as follows. The function $v(\rho)$ for all times t is the solution of the transformed elliptic problem

$$A(\rho)v = 0 \quad \text{in } \Omega_1^\Sigma \cup \Omega_2^\Sigma\,, \tag{4.68}$$

$$v = K(\rho) \quad \text{on } \Sigma\,, \tag{4.69}$$

$$\nabla v \cdot n = 0 \quad \text{on } \partial\Omega \tag{4.70}$$

where

$$A(\rho)v^i = (\Delta(v^i \circ F_\rho^{-1})) \circ F_\rho$$

for $v^i = v_{|\Omega_i^\Sigma}$ and $K(\rho)$ is the transformed mean curvature operator, i.e.,

$$K(\rho) := \kappa_\rho \circ F_\rho \text{ on } \Sigma\,,$$

where $\kappa_\rho(t,.)$ is the mean curvature of $\Gamma(t)$. It remains to define $B(\rho)$. With the help of

$$B^i(\rho)v^i := (\nabla(v^i \circ F_\rho^{-1}) \cdot \nabla\phi_\rho) \circ F_\rho$$

with $\phi_\rho(t,x) = d_\Sigma(t,x) - \rho(t, \pi_\Sigma(x))$, see Escher and Simonett [64] for details, we define

$$B(\rho)v := B^1(\rho)v^1 - B^2(\rho)v^2 \text{ on } \Sigma\,.$$

We notice that $u = v \circ F_\rho^{-1}$ is the unique solution of (4.63), (4.65) and (4.66) if and only if v solves (4.68)–(4.70). We are now left with finding a solution $\rho : [0,T] \times \Sigma \to (-a,a)$ for (4.67). In order to do so we need to compute $v(\rho)$ as a solution of (4.68)–(4.70), i.e., the problems (4.68)–(4.70) and (4.67) are coupled.

Remark 4.5.1

(1) For a complete local existence result we refer to [64, 129].
(2) In contrast to the mean curvature flow

$$V = \kappa$$

in the Mullins–Sekerka-problem the evolution depends in a non-local way upon the mean curvature.
(3) Other interface problems such as the Stefan problem and the two-phase flow problem can be transformed and solved in a similar fashion, see [129]. As Prüss and Simonett write: "The essential restriction is that the problem in question ought to be of *parabolic nature*".

Implicit Approaches for Interfaces 5

Abstract

In this chapter we discuss various implicit approaches in order to study the evolution of interfaces. Typically, these methods yield global solutions allowing for singularities in the flow. We present both the level set method and a BV-approach for obtaining global weak solutions of mean curvature flow. Another important implicit approach is given by phase-field models and we discuss existence, discretization and sharp-interface limits for the Cahn-Hilliard equation.

5.1 A Way to Handle Topological Changes: The Level Set Method

As we have seen in the previous chapter typically singularities in mean curvature flow occur, see also Fig. 5.1.

In such a case, provided a maximum comparison principle holds, the level set method can be used. As discussed before one can describe the interface as a level set of a scalar function as follows

$$\Gamma(t) = \{x \in I\!R^n \mid \phi(t, x) = 0\}\,.$$

Assuming $\nabla\phi(t, \cdot) \neq 0$ on $\Gamma(t)$ then a unit normal to $\Gamma(t)$ is given as

$$\nu(t, \cdot) = \frac{\nabla\phi(t, \cdot)}{|\nabla\phi(t, \cdot)|} \quad \text{on } \Gamma(t)\,.$$

E. Bänsch et al., *Interfaces: Modeling, Analysis, Numerics*,
Oberwolfach Seminars 51, https://doi.org/10.1007/978-3-031-35550-9_5

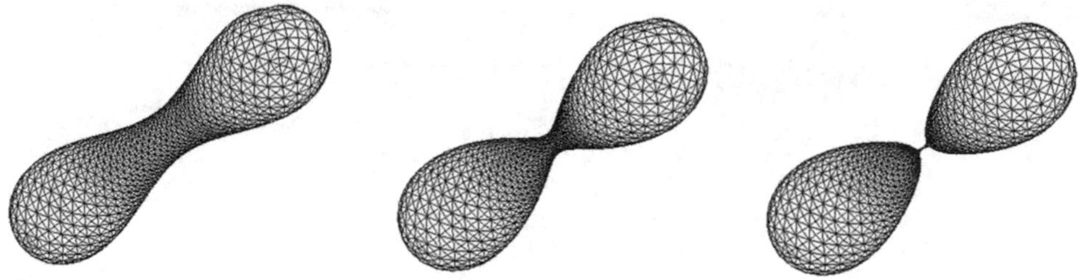

Fig. 5.1 Pinching singularity for interface evolution. Numerical simulation by Robert Nürnberg, see also [16]

Using Proposition 2.3.3 we obtain

$$\kappa(t,\cdot) = -\nabla\cdot\left(\frac{\nabla\phi(t,\cdot)}{|\nabla\phi(t,\cdot)|}\right) = -\frac{1}{|\nabla\phi(t,\cdot)|}\sum_{i,j=1}^{n}\left(\delta_{ij} - \frac{\partial_i\phi(t,\cdot)\partial_j\phi(t,\cdot)}{|\nabla\phi(t,\cdot)|^2}\right)\partial_{ij}\phi(t,\cdot)$$

$$= -\frac{1}{|\nabla\phi(t,\cdot)|}\left(\mathrm{Id} - \frac{\nabla\phi(t,\cdot)\otimes\nabla\phi(t,\cdot)}{|\nabla\phi(t,\cdot)|^2}\right) : D^2\phi(t,\cdot)\,. \tag{5.1}$$

Here, we define for matrices $A, B \in I\!R^{n\times n}$ the inner product

$$A : B = tr(A^T B) = \sum_{i,j=1}^{n} A_{ij}B_{ij}.$$

Next, in order to calculate the normal velocity $V(t_0, x_0)$ for some point $x_0 \in \Gamma(t_0)$ we choose as in Definition 2.7.1 a curve $\gamma : (t_0 - \delta, t_0 + \delta) \to I\!R^n$ with $\gamma(t) \in \Gamma(t)$ and $\gamma(t_0) = x_0$. Since $\phi(t, \gamma(t)) = 0,\ |t - t_0| < \delta$ we obtain

$$0 = \frac{d}{dt}\phi(t,\gamma(t))_{|t=t_0} = \partial_t\phi(t_0,x_0) + \nabla\phi(t_0,x_0)\cdot\gamma'(t_0)\,.$$

This implies

$$V(t_0,x_0) = \nu(t_0,x_0)\cdot\gamma'(t_0) = \frac{\nabla\phi(t_0,x_0)}{|\nabla\phi(t_0,x_0)|}\cdot\gamma'(t_0) = -\frac{\partial_t\phi(t_0,x_0)}{|\nabla\phi(t_0,x_0)|}\,. \tag{5.2}$$

Comparing (5.1) and (5.2) we see that the hypersurfaces $\Gamma(t)$ evolve according to $V = \kappa$ if ϕ is a solution of the equation

$$\partial_t\phi = \left(\mathrm{Id} - \frac{\nabla\phi\otimes\nabla\phi}{|\nabla\phi|^2}\right) : D^2\phi\,. \tag{5.3}$$

Defining $A(\nabla\phi) = \mathrm{Id} - \frac{\nabla\phi\otimes\nabla\phi}{|\nabla\phi|^2}$ we observe

$$A(\nabla\phi)\nabla\phi = 0$$

which implies that Eq. (5.3) is degenerate parabolic. This in particular means that the contraction mapping principle which has been used in the graph case, see Sect. 4.3.2, cannot be used. The fact that Eq. (5.3) degenerates in the direction $\nabla\phi$ is due to the fact that the evolution of each level set only depends on the level set itself and not on values normal to it where the normal direction is given by the direction $\nabla\phi$.

5.2 Viscosity Solutions for Mean Curvature Flow

We want to solve the following initial value problem:

$$\partial_t\phi - \left(\mathrm{Id} - \frac{\nabla\phi\otimes\nabla\phi}{|\nabla\phi|^2}\right) : D^2\phi = 0 \qquad \text{in } (0,\infty)\times I\!R^n\,, \tag{5.4}$$

$$\phi(0,x) = \phi_0(x) \quad \text{in } I\!R^n\,. \tag{5.5}$$

The notion of viscosity solution provides a powerful tool in order to solve highly nonlinear and possibly degenerate elliptic and parabolic partial differential equations of second order. We refer to the monograph [82] by Giga for an introduction to the level set approach for geometric evolution equations based on the theory of viscosity solutions. The corresponding analysis in the case of mean curvature flow was developed by Evans and Spruck [67] and Chen et al. [37].

We first try to motivate the following definition of a viscosity subsolution. Assume a smooth function ϕ fulfills

$$\partial_t\phi - \left(\mathrm{Id} - \frac{\nabla\phi\otimes\nabla\phi}{|\nabla\phi|^2}\right) : D^2\phi \le 0$$

in a point (t_0, x_0) with $\nabla\phi(t_0,x_0) \neq 0$. Suppose in addition that ψ is smooth and $\phi - \psi$ has a local maximum at (t_0, x_0). Then we obtain

$$\partial_t\phi(t_0,x_0) = \partial_t\psi(t_0,x_0),\;\; \nabla\phi(t_0,x_0) = \nabla\psi(t_0,x_0),\;\; D^2\phi(t_0,x_0) \le D^2\psi(t_0,x_0)\,.$$

We hence deduce (where the first inequality follows as in the proof of the classical maximum principle)

$$\begin{aligned}
&\partial_t \psi(t_0, x_0) - \left(\mathrm{Id} - \frac{\nabla\psi(t_0, x_0) \otimes \nabla\psi(t_0, x_0)}{|\nabla\psi(t_0, x_0)|^2} \right) : D^2\psi(t_0, x_0) \\
&\leq \partial_t \phi(t_0, x_0) - \left(\mathrm{Id} - \frac{\nabla\phi(t_0, x_0) \otimes \nabla\phi(t_0, x_0)}{|\nabla\phi(t_0, x_0)|^2} \right) : D^2\phi(t_0, x_0) \\
&\leq 0 .
\end{aligned}$$

Therefore we define viscosity solutions as follows.

Definition 5.2.1

(i) A function $\phi \in C^0([0, \infty) \times I\!R^n)$ is called a viscosity subsolution of (5.4) provided that for each $\psi \in C^\infty(I\!R^{n+1})$, for which $\phi - \psi$ has a local maximum at $(t_0, x_0) \in (0, \infty) \times I\!R^n$, we have

$$\begin{aligned}
\partial_t \psi - \left(\mathrm{Id} - \frac{\nabla\psi \otimes \nabla\psi}{|\nabla\psi|^2} \right) : D^2\psi \leq 0 \ \text{ at } (t_0, x_0) \text{ if } \nabla\psi(t_0, x_0) \neq 0 , \\
\partial_t \psi - (\mathrm{Id} - p \otimes p) : D^2\psi \leq 0 \ \text{ at } (t_0, x_0) \text{ for some } \\
|p| \leq 1 \ \text{ if } \nabla\psi(t_0, x_0) = 0 .
\end{aligned}$$

(ii) We define a viscosity supersolution of (5.4) analogously by replacing maximum by minimum and demanding

$$\begin{aligned}
\partial_t \psi - \left(\mathrm{Id} - \frac{\nabla\psi \otimes \nabla\psi}{|\nabla\psi|^2} \right) : D^2\psi \geq 0 \ \text{ at } (t_0, x_0) \text{ if } \nabla\psi(t_0, x_0) \neq 0 , \\
\partial_t \psi - (\mathrm{Id} - p \otimes p) : D^2\psi \geq 0 \ \text{ at } (t_0, x_0) \text{ for some } \\
|p| \leq 1 \ \text{ if } \nabla\psi(t_0, x_0) = 0 .
\end{aligned}$$

(iii) A viscosity solution of (5.4), (5.5) is a function which is both a sub- and a supersolution and which satisfies $\phi(0, x) = \phi_0(x)$ for all $x \in I\!R^n$.

5.3 An Existence Theorem for Viscosity Solutions of Mean Curvature Flow

Evans and Spruck [67] and Chen et al. [37] proved the following theorem.

Theorem 5.3.1 *Assume $\phi_0 : I\!R^n \to I\!R$ is continuous and satisfies*

$$\phi_0(x) = 1 \textit{ for } |x| \geq S$$

for some $S > 0$. Then there exists a unique viscosity solution of (5.4), (5.5) *such that*

$$\phi(t, x) = 1 \textit{ for } |x| + t \geq R$$

for some $R > 0$ depending only on S.

Remark 5.3.2

(1) Given a compact hypersurface Γ_0 we can choose a continuous function $\phi_0 : I\!R^n \to I\!R$ such that

$$\Gamma_0 = \{x \in I\!R^n \mid \phi_0(x) = 0\}.$$

If $\phi : [0, \infty) \times I\!R^n \to I\!R$ is the unique viscosity solution of (5.4), (5.5) we then call

$$\Gamma(t) = \{x \in I\!R^n \mid \phi(t, x) = 0\}, \ t \geq 0$$

a generalized evolution by mean curvature flow. One can show that the sets $\Gamma(t)$ only depend on Γ_0 but not on the specific choice of the level set function ϕ_0.

(2) The sets $(\Gamma(t))_{t \geq 0}$ exist for all times and lead to a notion of a solution past singularities.

(3) The sets $\Gamma(t)$ can have an interior. This is called fattening and in this case we do not obtain the evolution of a hypersurface, see Fig. 5.2 for an example and the book of Giga [82] for more details.

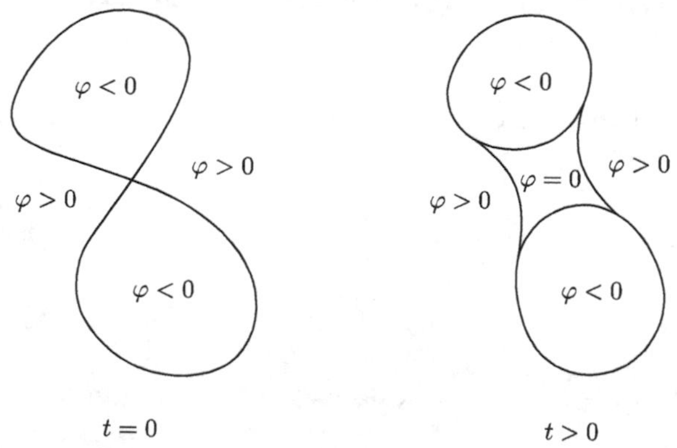

Fig. 5.2 An example for fattening in mean curvature flow

Evans and Spruck [67] used a regularized problem to show the existence result in Theorem 5.3.1. They solved

$$\partial_t \phi^\varepsilon - \left(\mathrm{Id} - \frac{\nabla\phi^\varepsilon \otimes \nabla\phi^\varepsilon}{\varepsilon^2 + |\nabla\phi^\varepsilon|^2}\right) : D^2\phi^\varepsilon = 0 \quad \text{in } (0,\infty) \times I\!R^n, \tag{5.6}$$

$$\phi^\varepsilon(0,\cdot) = \phi_0 \quad \text{in } I\!R^n \tag{5.7}$$

and showed that the sequence $(\phi^\varepsilon)_{\varepsilon>0}$ converges locally uniformly in $[0,\infty) \times I\!R^n$ to a viscosity solution ϕ of (5.4), (5.5). In the proof, maximum and comparison principles are used in crucial arguments.

5.4 A Level Set Approach for Numerically Solving Mean Curvature Flow

We present an algorithm based on the level-set formulation for the mean curvature flow from Sect. 5.2, see [72]. In order to handle potential difficulties when $\nabla\phi$ vanishes we use the regularized problem (5.6) which we rewrite similarly as in Sect. 5.1 as

$$\begin{aligned} 0 &= \partial_t \phi^\varepsilon - \left(\mathrm{Id} - \frac{\nabla\phi^\varepsilon \otimes \nabla\phi^\varepsilon}{\varepsilon^2 + |\nabla\phi^\varepsilon|^2}\right) : D^2\phi^\varepsilon \\ &= \partial_t \phi^\varepsilon - \sqrt{\varepsilon^2 + |\nabla\phi^\varepsilon|^2}\,\nabla\cdot\Big(\frac{\nabla\phi^\varepsilon}{\sqrt{\varepsilon^2 + |\nabla\phi^\varepsilon|^2}}\Big). \end{aligned}$$

Unfortunately, the spatial operator is not in divergence form, which makes a finite element discretization difficult. Thus we divide by $\sqrt{\varepsilon^2 + |\nabla \phi^\varepsilon|^2}$ to arrive at

$$\frac{\partial_t \phi^\varepsilon}{\sqrt{\varepsilon^2 + |\nabla \phi^\varepsilon|^2}} - \nabla \cdot \Big(\frac{\nabla \phi^\varepsilon}{\sqrt{\varepsilon^2 + |\nabla \phi^\varepsilon|^2}} \Big) = 0. \tag{5.8}$$

For practical purposes it is necessary to solve (5.8) on a bounded computational domain. Therefore we choose a bounded convex domain $\Omega \subset I\!R^n$ that contains the initial hypersurface Γ_0 and consider (5.8) together with a homogeneous Neumann boundary condition. The discretization is now straightforward: The time derivative is replaced by, for instance, a backward Euler time discretization and the spatial operator is discretized by integrating by parts and then using finite elements. More precisely, let $\mathscr{T}$ be a triangulation of $\overline{\Omega}$ and denote by

$$X_h := \{\phi_h \in C^0(\overline{\Omega}) \,|\, \phi_{h|T} \in P_1(T),\ T \in \mathscr{T}\}$$

the space of continuous, piecewise linear finite elements. Furthermore, we choose a time step $\delta > 0$ and set $t_k = k\delta, k = 0, \ldots, M$ with $M\delta = T$. We denote by $\phi_h^k \in X_h$ the approximation of $\phi^\varepsilon(t_k, \cdot)$ and choose $\phi_h^0 = I_h \phi_0$, where I_h is the Lagrange interpolation operator and $\phi_0 \in C^0(\overline{\Omega})$ such that $\Gamma_0 = \{x \in I\!R^n \,|\, \phi_0(x) = 0\}$, compare (1) in Remark 5.3.2.

The algorithm is now given as follows: Given $\phi_h^k \in X_h$ find $\phi_h^{k+1} \in X_h$ such that

$$\frac{1}{\delta} \int_\Omega \frac{(\phi_h^{k+1} - \phi_h^k)\zeta_h}{\sqrt{\varepsilon^2 + |\nabla \phi_h^k|^2}}\, dx + \int_\Omega \frac{\nabla \phi_h^{k+1} \cdot \nabla \zeta_h}{\sqrt{\varepsilon^2 + |\nabla \phi_h^k|^2}}\, dx = 0 \qquad \forall \zeta_h \in X_h. \tag{5.9}$$

Note that in each time step only a linear system has to be solved. Somewhat surprisingly the scheme is nevertheless unconditionally stable in the sense that

$$\int_\Omega \sqrt{\varepsilon^2 + |\nabla \phi_h^k|^2}\, dx \le \int_\Omega \sqrt{\varepsilon^2 + |\nabla \phi_h^0|^2}\, dx, \quad k = 1, \ldots, M. \tag{5.10}$$

In order to see (5.10) we adapt the proof of Theorem 5.5 in [46]. Inserting $\zeta_h = \phi_h^{k+1} - \phi_h^k$ into (5.9) we obtain

$$\frac{1}{\delta} \int_\Omega \frac{(\phi_h^{k+1} - \phi_h^k)^2}{\sqrt{\varepsilon^2 + |\nabla \phi_h^k|^2}}\, dx + \int_\Omega \frac{\nabla \phi_h^{k+1} \cdot \nabla(\phi_h^{k+1} - \phi_h^k)}{\sqrt{\varepsilon^2 + |\nabla \phi_h^k|^2}}\, dx = 0. \tag{5.11}$$

Let us focus on the second term on the left hand side and write

$$A := \frac{\nabla\phi_h^{k+1}\cdot\nabla(\phi_h^{k+1}-\phi_h^k)}{\sqrt{\varepsilon^2+|\nabla\phi_h^k|^2}} = \frac{|\nabla\phi_h^{k+1}|^2}{\sqrt{\varepsilon^2+|\nabla\phi_h^k|^2}} - \frac{\nabla\phi_h^{k+1}\cdot\nabla\phi_k^k}{\sqrt{\varepsilon^2+|\nabla\phi_h^k|^2}}.$$

Abbreviating $Q_h^k = \sqrt{\varepsilon^2+|\nabla\phi_h^k|^2}$ and $\nu_h^k = \frac{(\nabla\phi_h^k,\varepsilon)^T}{Q_h^k}$ we see that

$$|\nu_h^k| = 1 \quad \text{as well as} \quad \nu_h^{k+1}\cdot\nu_h^k = \frac{\nabla\phi_h^{k+1}\cdot\nabla\phi_h^k+\varepsilon^2}{Q_h^{k+1}Q_h^k},$$

so that after some straightforward calculations

$$\begin{aligned} A &= \frac{(Q_h^{k+1})^2-\varepsilon^2}{Q_h^k} - Q_h^{k+1}\,\nu_h^{k+1}\cdot\nu_h^k + \frac{\varepsilon^2}{Q_h^k} = \frac{(Q_h^{k+1})^2}{Q_h^k} - Q_h^{k+1}\,\nu_h^{k+1}\cdot\nu_h^k \\ &= Q_h^{k+1} - Q_h^k + \frac{1}{2}|\nu_h^{k+1}-\nu_h^k|^2 Q_h^{k+1} + \frac{(Q_h^{k+1}-Q_h^k)^2}{Q_h^k} \\ &\geq Q_h^{k+1} - Q_h^k. \end{aligned}$$

If we insert this inequality into (5.11) and recall the definition of Q_h^k we obtain

$$\int_\Omega \sqrt{\varepsilon^2+|\nabla\phi_h^{k+1}|^2}\,dx \leq \int_\Omega \sqrt{\varepsilon^2+|\nabla\phi_h^k|^2}\,dx,$$

so that (5.10) follows. A convergence result for the Algorithm (5.9) can be found in [43]. Fig. 5.3 shows a simulation, where the initial curve is chosen as a lemniscate. In Fig. 5.4 we present a numerical computation of mean curvature flow with the level set method in three dimensions where the initial zero level set is a torus.

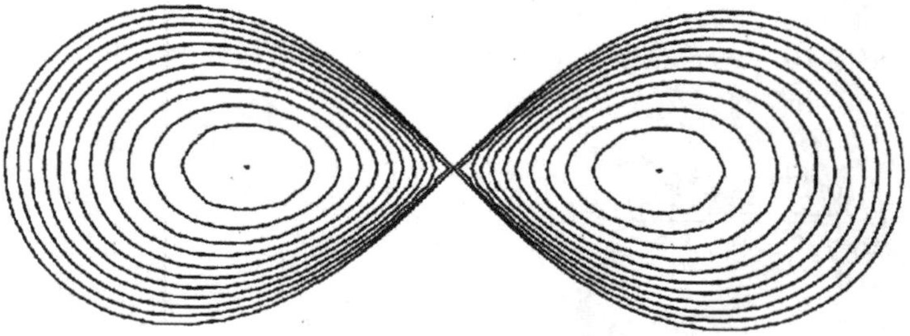

Fig. 5.3 Curvature flow for curves computed with the level set method; picture taken from [73]

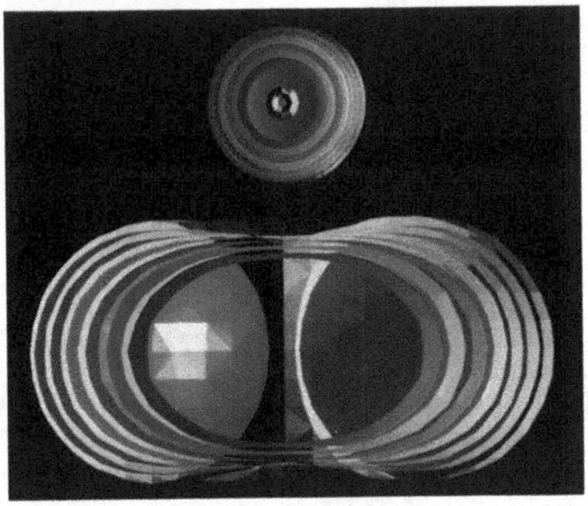

Fig. 5.4 Mean curvature flow computed with the level set method: evolution of the zero level set for a 3D situation. The initial zero level set is a torus, evolving towards a sphere, thus changing its topology. Cut perpendicular to the plane of rotational symmetry (below), cut parallel to this plane (above). Picture taken from [73]

5.5 Relating Phase Field and Sharp Interface Energies

The Ginzburg–Landau energy

$$E_\varepsilon(\varphi) := \int_\Omega (\tfrac{\varepsilon}{2}|\nabla\varphi|^2 + \tfrac{1}{\varepsilon}\psi(\varphi))dx, \tag{5.12}$$

can be related to the surface energy in the limit $\varepsilon \to 0$. The appropriate notion to make this statement precise is the concept of Γ-limit which we now define.

Definition 5.5.1 Let (X, d) be a metric space and $(F_\varepsilon)_{\varepsilon>0}$ a family of functionals $F_\varepsilon : X \to (-\infty, \infty]$. We say that $(F_\varepsilon)_{\varepsilon>0}$ Γ-converges to a functional $F : X \to (-\infty, \infty]$ (which we will denote as $F_\varepsilon \overset{\Gamma}{\longrightarrow} F$) if the following properties hold:

(i) (*lim inf inequality*) For every $u \in X$ and $u_\varepsilon \in X$, $\varepsilon > 0$, such that $u_\varepsilon \to u$ as $\varepsilon \to 0$ it holds

$$F(u) \le \liminf_{\varepsilon\to 0} F_\varepsilon(u_\varepsilon)\,.$$

(ii) (*lim sup inequality*) For every $u \in X$ there exist $u_\varepsilon \in X$, $\varepsilon > 0$, such that $u_\varepsilon \to u$ as $\varepsilon \to 0$ and

$$\limsup_{\varepsilon\to 0} F_\varepsilon(u_\varepsilon) \le F(u)\,.$$

We note that the concept of Γ-limits is more general and can be generalized to more general spaces, see Dal Maso [40]. The notion of Γ-limit is in particular appropriate for sequences of variational problems as under appropriate assumptions minima of F_ε will converge to minima of F, see [31].

It was shown in [120] and [121] that the Ginzburg–Landau energies E_ε defined in Eq. (5.12) Γ-converge to a multiple of the surface area functional. It turns out that a suitable metric for this convergence is induced by the $L^1(\Omega)$-norm and hence we extend E_ε to $L^1(\Omega)$ by setting

$$E_\varepsilon(\varphi) := \begin{cases} \int_\Omega (\frac{\varepsilon}{2}|\nabla\varphi|^2 + \frac{1}{\varepsilon}\psi(\varphi))\, dx & \text{if } \varphi \in H^1(\Omega)\,, \\ \infty & \text{if } \varphi \in L^1(\Omega) \setminus H^1(\Omega)\,. \end{cases}$$

In order to formulate this precisely we work in the space of bounded variation (BV). Define $f \in BV(\Omega)$, if $f \in L^1(\Omega)$ and $\int_\Omega |\nabla f|$, the total variation of the distribution ∇f, is finite, i.e.,

$$\int_\Omega |\nabla f| := \sup\left\{ \int_\Omega f\; \nabla\cdot g\, dx \mid g \in C_0^1(\Omega, I\!R^n),\ |g(x)| \le 1 \text{ for all } x \in \Omega \right\} < \infty.$$

For $f \in BV(\Omega)$ one obtains that ∇f and $|\nabla f|$ are Radon measures on Ω with values in $I\!R^n$ and $I\!R$, respectively. A measurable set $E \subset \Omega$ with $\int_\Omega |\nabla \chi_E| < \infty$, where χ_E is the characteristic function of E, is called Caccioppoli set. In a generalized sense such a set E has a bounded perimeter. We can now define a generalized unit normal to the boundary of E given by $\nu_E = \frac{\nabla \chi_E}{|\nabla \chi_E|}$ as the Radon–Nikodym derivative of $\nabla \chi_E$ with respect to $|\nabla \chi_E|$. We refer to Giusti [84] and Ambrosio, Fusco, Pallara [7] for more details on functions of bounded variation.

If $E \subset\subset \Omega$ is open with smooth boundary it holds

$$\int_\Omega |\nabla \chi_E| = \mathscr{H}^{n-1}(\partial E)\,. \tag{5.13}$$

To show this is an exercise. Under appropriate assumptions on ψ and Ω it can be shown that the functionals E_ε in fact Γ-converge to the functional

$$E(\varphi) := \begin{cases} c_\psi \int_\Omega |\nabla \chi_{\{\varphi=1\}}| & \text{if } \varphi \in BV(\Omega, \{-1, 1\})\,, \\ \infty & \text{if } \varphi \in L^1(\Omega) \setminus BV(\Omega, \{-1, 1\})\,, \end{cases}$$

where $c_\psi := \int_{-1}^1 \sqrt{2\psi(z)}dz$. More precisely, we have

$$E_\varepsilon \xrightarrow{\Gamma} E \quad \text{as} \quad \varepsilon \to 0$$

with respect to the L^1-topology. This Γ-convergence result is stable under adding an integral constraint for E_ε in the functional which is important in many applications where this corresponds to a mass conservation property. We refer to [120, 121] and [31] for more details. It is possible to relate gradient flows of E_ε to the gradient flows of the area functional E discussed in Sect. 3.3, see [76].

5.6 Solving Interface Evolution Problems in a BV-Setting

In general classical solutions to interface evolution problems do not exist for large times due to the fact that topological changes and singularities can occur. We already introduced viscosity solutions as a way to have a weak formulation allowing for singularities in the geometry. Another approach for long-time existence allowing for singularities is a setting within functions of bounded variations (BV-functions). This has been used for example by Luckhaus and Sturzenhecker [114], see also [5] for a related approach. Luckhaus and Sturzenhecker [114], see also [113], used a weak formulation of the identity $u = \kappa$ in the setting of functions of bounded variations (BV-functions). The BV-formulation of Luckhaus and Sturzenhecker [114] now replaces the pointwise identity $u = \kappa$ by

$$\int_0^T \int_\Omega \left(\nabla \cdot \xi - \frac{\nabla \chi}{|\nabla \chi|} \cdot \left(D\xi \frac{\nabla \chi}{|\nabla \chi|} \right) \right) d|\nabla \chi(t)| dt = \int_{\Omega_T} \nabla \cdot (u\xi)\, \chi \, d(t,x)\,, \qquad (5.14)$$

which has to hold for all $\xi \in C^1(\overline{\Omega}_T, I\!R^n)$, $\Omega_T := (0,T) \times \Omega$. Here $\chi : \Omega_T \to \{0,1\}$ is a phase function where phase 2 is given by the set $\{(t,x) \in (0,T) \times \Omega \mid \chi(t,x) = 1\}$ and phase 1 is given by the set $\{(t,x) \in (0,T) \times \Omega \mid \chi(t,x) = 0\}$ and one assumes that $\chi(t,.) \in BV(\Omega)$ for all $t \in (0,T)$.

If the interface is smooth and without boundary Eq. (5.14) leads to

$$\int_0^T \int_{\Gamma(t)} \nabla_\Gamma \cdot \xi \, d\mathcal{H}^{n-1} dt = - \int_0^T \int_{\Gamma(t)} u\, \xi \cdot \nu \, d\mathcal{H}^{n-1} dt$$

and using the Gauss theorem on manifolds we have

$$\int_0^T \int_{\Gamma(t)} \kappa \xi \cdot \nu \, d\mathcal{H}^{n-1} dt = \int_0^T \int_{\Gamma(t)} u\, \xi \cdot \nu \, d\mathcal{H}^{n-1} dt$$

which shows that (5.14) is a weak formulation of $u = \kappa$.

Luckhaus and Sturzenhecker [114] use the method of implicit time discretization in order to approximate solutions to the mean curvature flow equation and to the Mullins–Sekerka problem. This is done in the spirit of a minimizing movement scheme discussed in Sect. 3.1.2. The discretization for mean curvature flow is done as follows. For $\delta = T/N$, $N \in \mathbb{N}$, a time-discrete solution $\chi_\delta : \Omega_T \to \{0,1\} \in L^1(0,T; BV(\Omega))$ is constructed which is constant on intervals $[(i-1)\delta, i\delta)$. The construction is done iteratively by minimizing the energy functional

$$\mathscr{F}(E) = \int_\Omega |\nabla \chi_E| + \int_{E \Delta E_{t-\delta}} \left(\tfrac{1}{\delta}\right) \operatorname{dist}\,(.,\partial E_{t-\delta})\, dx \tag{5.15}$$

in the class of all measurable subsets $E \subset \Omega$, where we start on $[0, \delta)$ with the set E_0. The sets E_t are then defined as a minimizer of the above minimization problem. The notation $E \Delta F$ stands for the symmetric difference of two sets

$$E \Delta F = (E \setminus F) \cup (F \setminus E)\,.$$

The term $\int_{E \Delta E_{t-\delta}} (\frac{1}{\delta}) \operatorname{dist}\,(.,\partial E_{t-\delta})$ in (5.15) is quadratic in the "distance" of the sets E and $E_{t-\delta}$ (take into account that an integration in normal direction is involved). One can show existence of time-discrete solutions and a compactness result yields $\chi_\delta \to \chi$ in $L^1(\Omega_T)$ as $\delta \to 0$ for some $\chi \in L^2(\Omega_T)$. It is shown in addition that the $\chi_\delta(t) \in BV(\Omega)$ are uniformly bounded in t. One also obtains a $u : \Omega_T \to I\!R$ with $u \in L^1(0, T; L^1(|\nabla \chi(t)|)$ (careful: interpret this in the correct way) such that

$$\int_{\Omega_T} \left(\left(\nabla \cdot \xi - \frac{\nabla \chi}{|\nabla \chi|} \cdot \left(\nabla \xi \frac{\nabla \chi}{|\nabla \chi|} \right) \right) |\nabla \chi| + u \xi \cdot \nabla \chi \right) = 0 \tag{5.16}$$

for all $\xi \in C^\infty(\overline{\Omega_T}, I\!R^n)$, $\xi_{|(0,T)\times\partial\Omega} = 0$ and

$$\int_{\Omega_T} \chi \partial_t \zeta + \int_\Omega \chi_{E_0} \zeta(0) = - \int_{\Omega_T} u \zeta |\nabla \chi|$$

for all $\zeta \in C^\infty(\overline{\Omega_T}, I\!R)$, $\zeta_{|(0,T)\times\partial\Omega} = 0$, $\zeta(T) = 0$.

The first equation states

$$\kappa = u$$

and the second is a weak form of

$$V = u\,.$$

However, in order to ensure that the limit procedure can be performed rigorously situations like in Fig. 5.5 have to be excluded. This approach can also be used for situations when an equation on Γ is coupled to bulk equations. A simplified form of the Stefan problem is

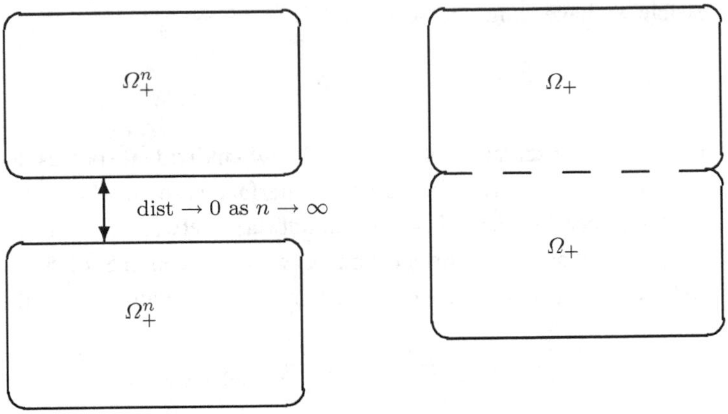

Fig. 5.5 An example where a loss of area appears in a limit $n \to \infty$

the following Mullins–Sekerka-problem

$$-\Delta u = 0 \quad \text{in } \Omega_-(t) \cup \Omega_+(t)\,, \tag{5.17}$$

$$V = -[\nabla u]_-^+ \cdot \nu \quad \text{on } \Gamma(t)\,, \tag{5.18}$$

$$u = \kappa \quad \text{on } \Gamma(t)\,, \tag{5.19}$$

$$\nabla u \cdot n = 0 \quad \text{on } \partial\Omega\,, \tag{5.20}$$

where $\Omega_-(t)$, $\Omega_+(t)$ are the sets occupied by the two phases at time t. In this case (5.16) is coupled to the following weak formulation of (5.17), (5.18) and (5.19): Find a function $u \in L^2((0,T), H^{1,2}(\Omega))$ such that

$$\int_{\Omega_T} \chi\, \partial_t \zeta + \int_\Omega \chi_{E_0} \zeta(0) = \int_{\Omega_T} \nabla u \cdot \nabla \zeta$$

for all $\zeta \in C^\infty(\overline{\Omega_T}, I\!R)$, $\zeta(T) = 0$. For a related result for the Stefan problem we refer to Luckhaus [113].

5.7 Phase Field Models for Two-Phase Flow: The Cahn–Hilliard–Navier–Stokes Model

In this section we consider a phase field description of the two-phase flow problem introduced in Sect. 3.7. In a phase field model of two-phase flow, a partial mixing of the two incompressible fluids in a thin interfacial region is assumed. At the beginning we consider for simplicity the case where the two fluids have the same density. As the fluids are incompressible we have, compare Sect. 3.7.1,

$$\nabla \cdot u = 0. \tag{5.21}$$

This equation gives the conservation of mass for the two individual species. However, the conservation of momentum derived in the sharp interface case, see Sect. 3.7.2 needs to be adapted. We have seen in Sect. 3.7.4 that the interface between two phases carries an energy which is proportional to the surface area. As we have seen in Sect. 5.5, surface area in the phase field setting is approximated by the Ginzburg–Landau functional

$$E_\varepsilon(\varphi) := \int_\Omega (\tfrac{\varepsilon}{2}|\nabla\varphi|^2 + \tfrac{1}{\varepsilon}\psi(\varphi))dx. \tag{5.22}$$

Motivated by this and by Proposition 3.7.2 we introduce a total energy density

$$e(u, \varphi, \nabla\varphi) = \frac{\rho}{2}|u|^2 + f(\varphi, \nabla\varphi) \tag{5.23}$$

as the sum of the kinetic energy and an interfacial energy. We take

$$f(\varphi, \nabla\varphi) = \hat{\gamma}(\tfrac{\varepsilon}{2}|\nabla\varphi|^2 + \tfrac{1}{\varepsilon}\psi(\varphi)) \tag{5.24}$$

where $\hat{\gamma}$ is a parameter proportional to the coefficient of surface tension γ. Similar as in Chap. 3 one can derive a balance equation for the phase field as

$$\partial_t\varphi + \nabla \cdot (\varphi u) + \nabla \cdot J_\varphi = 0 \tag{5.25}$$

and a momentum equation as

$$\rho\partial_t u + \rho u \cdot \nabla u = \nabla \cdot \mathbf{T}. \tag{5.26}$$

The appropriate formulation of the second law of thermodynamics in the isothermal case is given by the following dissipation inequality, see e.g. [89],

$$\frac{d}{dt}\int_{V(t)} e(u, \varphi, \nabla\varphi)\,dx + \int_{\partial V(t)} J_e \cdot \nu_S\, d\mathcal{H}^{n-1} \leq 0 \tag{5.27}$$

where $V(t)$ is a test volume which is transported with the flow u, J_e is an energy flux and ν_S the outward pointing normal of $\partial V(t)$. Using the Transport Theorem 2.11.1 and the fact that the test volume is arbitrary one obtains the local form, see [3, 111],

$$-\mathcal{D} := \partial_t e + \nabla \cdot (ue) + \nabla \cdot J_e \leq 0. \tag{5.28}$$

One can now use a Lagrange multiplier method, see [111], to derive constitutive relations between the different involved quantities which guarantee that the second law is fulfilled. Every fields (φ, u) which fulfill the dissipation inequality (5.28) and $\nabla \cdot u = 0$ also fulfill

$$-\mathscr{D} = \partial_t e + u \cdot \nabla\varphi + \nabla \cdot J_\varphi - \mu(\partial_t \varphi + u \cdot \nabla\varphi + \nabla \cdot J_\varphi) \leq 0, \tag{5.29}$$

where μ is a Lagrange multiplier which will be specified later.

Using the fact that the velocity is divergence free one obtains

$$\begin{aligned}\partial_t \left(\frac{\rho}{2}|u|^2\right) + \nabla \cdot \left(\frac{\rho}{2}|u|^2 u\right) &= (\nabla \cdot \mathbf{T}) \cdot u \\ &= \nabla \cdot \left(\mathbf{T}^T u\right) - \mathbf{T} : \nabla u \,.\end{aligned}$$

Denoting by $f_{,\varphi}$ and $f_{,\nabla\varphi}$ the partial derivatives with respect to φ and $\nabla\varphi$ one gets

$$D_t f = f_{,\varphi} D_t \varphi + f_{,\nabla\varphi} \cdot D_t \nabla\varphi$$

where

$$D_t u = \partial_t u + u \cdot \nabla u$$

is the material time derivative. Using

$$D_t \nabla\varphi = \nabla D_t \varphi - (\nabla u)^T \nabla\varphi \tag{5.30}$$

yields that (5.29) gives after some computations

$$\begin{aligned}-\mathscr{D} = \nabla \cdot &\left(J_e + \mathbf{T}^T u - \mu J_\varphi + f_{,\nabla\varphi} D_t \varphi\right) \\ &+ (f_{,\varphi} - \mu - \nabla \cdot f_{,\nabla\varphi}) D_t \varphi \\ &- (\mathbf{T} + \nabla\varphi \otimes f_{,\nabla\varphi}) : \nabla u + \nabla\mu \cdot J_\varphi \leq 0 \,.\end{aligned}$$

Choosing the chemical potential as

$$\mu = f_{,\varphi} - \nabla \cdot f_{,\nabla\varphi}$$

and the energy flux as

$$J_e = -\mathbf{T}^T u + \mu J_\varphi - f_{,\nabla\varphi} D_t \varphi$$

one ends up with the dissipation inequality

$$(\mathbf{T} + \nabla\varphi \otimes f_{,\nabla\varphi}) : \nabla u - \nabla\mu \cdot J_\varphi \geq 0\,.$$

As in [89] we introduce an extra stress $\mathbf{S}$ and the pressure p such that

$$\widetilde{\mathbf{S}} = \mathbf{T} + p \ \mathrm{Id}\,.$$

Due to the incompressibility condition $\nabla \cdot u = 0$ the pressure p is still indeterminate, see also [89]. With the stress $\widetilde{\mathbf{S}}$ one obtains

$$(\widetilde{\mathbf{S}} + \nabla\varphi \otimes f_{,\nabla\varphi}) : \nabla u - \nabla\mu \cdot J_\varphi \geq 0$$

since $\nabla \cdot u = 0$. The term $\mathbf{S} = \widetilde{\mathbf{S}} + \nabla\varphi \otimes f_{,\nabla\varphi}$ will turn out to be the viscous stress tensor since it corresponds to irreversible changes of energy due to friction.

We now choose specific constitutive assumptions. In order to obtain a so-called Newtonian fluid we choose

$$\mathbf{S} = \widetilde{\mathbf{S}} + \nabla\varphi \otimes f_{,\nabla\varphi} = 2\hat{\mu}(\varphi)D(u)$$

with a φ-dependent viscosity $\hat{\mu}(\varphi) \geq 0$ and $D(u) = \frac{1}{2}\left(\nabla u + (\nabla u)^T\right)$. For the diffuse flux J_φ we choose a variant which is of Fick's type as follows

$$J = -m(\varphi)\nabla\mu,$$

where $m(\varphi) \geq 0$, which guarantees that the dissipation inequality is fulfilled. With the free energy

$$f(\varphi, \nabla\varphi) = \hat{\gamma}\left(\frac{\varepsilon}{2}|\nabla\varphi|^2 + \frac{1}{\varepsilon}\psi(\varphi)\right)$$

we obtain the following Cahn–Hilliard–Navier–Stokes model with matched densities

$$\rho\partial_t u + ((\rho u) \cdot \nabla)u - \nabla \cdot (2\hat{\mu}(\varphi)D(u)) + \nabla p = -\hat{\gamma}\varepsilon\nabla \cdot (\nabla\varphi \otimes \nabla\varphi)\,, \tag{5.31}$$

$$\nabla \cdot u = 0\,, \tag{5.32}$$

$$\partial_t\varphi + u \cdot \nabla\varphi = \nabla \cdot (m(\varphi)\nabla\mu)\,, \tag{5.33}$$

$$\frac{\hat{\gamma}}{\varepsilon}\psi'(\varphi) - \hat{\gamma}\varepsilon\Delta\varphi = \mu\,. \tag{5.34}$$

Matched densities means that the densities in the two phases are the same. This is of course seldom the case in real applications. Therefore, several attempts have been made to introduce Cahn–Hilliard–Navier–Stokes models with non-matched densities. The most prominent examples are models by Lowengrub and Truskinovsky [112] and Abels et al. [3]. In the model of Abels et al. the momentum balance (5.31) is replaced by

$$\rho\partial_t u + ((\rho u + \widetilde{J}) \cdot \nabla)u - \nabla \cdot (2\hat{\mu}(\varphi)D(u)) + \nabla p = -\hat{\gamma}\varepsilon\nabla \cdot (\nabla\varphi \otimes \nabla\varphi)\,,$$

with

$$\widetilde{J} = \frac{\tilde{\rho}_+ - \tilde{\rho}_-}{2} J_\varphi = -\frac{\tilde{\rho}_+ - \tilde{\rho}_-}{2} m(\varphi)\nabla\mu$$

where $\tilde{\rho}_+$ and $\tilde{\rho}_-$ are the mass densities of the two phases.

It can be shown, with the help of formally matched asymptotic expansions, that the above Cahn–Hilliard–Navier–Stokes system converges to the sharp interface problem introduced in Sect. 3.7, see [3]. A mathematical analysis of Cahn–Hilliard–Navier–Stokes system can be found in [1,4].

5.8 Existence Theory for the Cahn–Hilliard Equation

In this section we discuss how to show existence of solutions to the Cahn–Hilliard equation. The idea is to show in a simple situation how one can obtain existence results. For more complex situations we refer to [2, 4, 62, 119]. We consider the Cahn–Hilliard equation in the form

$$\partial_t\varphi = \Delta\mu\,, \tag{5.35}$$

$$\mu = -\varepsilon\Delta\varphi + \tfrac{1}{\varepsilon}\psi'(\varphi) \tag{5.36}$$

in $\Omega_T := (0, T) \times \Omega$. We assume that $\Omega \subset \mathbb{R}^n$, $n \in \mathbb{N}$, is a bounded domain with Lipschitz boundary and choose a time $T > 0$. Here $\varphi, \mu : (0, T) \times \Omega \to \mathbb{R}$ are the scaled concentration and the chemical potential. The parameter $\varepsilon > 0$ is a typically small constant and ψ is a free energy density and in applications ψ often has a double well form, see Sect. 3.8 for details. Equations (5.35) and (5.36) have to be solved together with Neumann and no-flux boundary conditions

$$\nabla\varphi \cdot n = 0 \quad \text{and} \quad \nabla\mu \cdot n = 0 \quad \text{on} \quad (\partial\Omega)_T := (0, T) \times \partial\Omega \tag{5.37}$$

and initial conditions $\varphi(0) = \varphi_0$. Here n is the outer unit normal to $\partial\Omega$. We assume

(A) that $\psi \in C^1(\mathbb{R}, \mathbb{R})$ and that there exist constants $C_1, C_2, C_3 > 0$ such that for all $z \in \mathbb{R}$

$$|\psi'(z)| \leq C_1 |z|^q + C_2 \quad \text{and} \quad \psi(z) \geq -C_3 ,$$

where $q = \frac{n}{n-2}$ if $n \geq 3$ and $q \in \mathbb{R}^+$ arbitrary if $n = 1, 2$.

With these assumptions we will prove an existence result. The proof of the existence result will crucially depend on the following energy decreasing property of solutions

$$\begin{aligned}
\frac{\mathrm{d}}{\mathrm{d}t} \int_\Omega \left(\tfrac{\varepsilon}{2} |\nabla\varphi|^2 + \tfrac{1}{\varepsilon}\psi(\varphi) \right) dx &= \int_\Omega \left(\varepsilon \nabla\varphi \cdot \nabla\partial_t\varphi + \tfrac{1}{\varepsilon}\psi'(\varphi)\partial_t\varphi \right) dx \\
&= \int_\Omega \partial_t\varphi(-\varepsilon\Delta\varphi + \tfrac{1}{\varepsilon}\psi'(\varphi)) dx \\
&= \int_\Omega (\Delta\mu)\mu \, dx \\
&= -\int_\Omega |\nabla\mu|^2 dx \leq 0 . \qquad (5.38)
\end{aligned}$$

Integrating this identity will give the basic a priori estimates for (5.35), (5.36), (5.37).

We now formulate the basic existence result for (5.35)–(5.37). For notations and results on Sobolev spaces and Banach space valued integration we refer to Alt [6], Evans [66] and Wloka [152].

Theorem 5.8.1 *We assume (A) and $\varphi_0 \in H^1(\Omega)$. Then there exists a pair of functions (φ, μ) such that*

(1) $\varphi \in L^\infty(0, T; H^1(\Omega)) \cap C^0([0, T]; L^2(\Omega))$,

(2) $\partial_t\varphi \in L^2(0, T; (H^1(\Omega))')$,

(3) $\varphi(0) = \varphi_0$,

(4) $\mu \in L^2(0, T; H^1(\Omega))$,

which satisfies (5.35) *and* (5.36) *in the following weak sense. It holds that*

$$\int_0^T \langle \partial_t\varphi(t), \zeta(t) \rangle_{(H^1)', H^1} \, dt = -\int_{\Omega_T} \nabla\mu \cdot \nabla\zeta \, d(t, x) \qquad (5.39)$$

(continued)

for all $\zeta \in L^2(0, T; H^1(\Omega))$ *and*

$$\int_{\Omega_T} \mu\eta \, d(t, x) = \int_{\Omega_T} \varepsilon \nabla\varphi \cdot \nabla\eta \, d(t, x) + \int_{\Omega_T} \tfrac{1}{\varepsilon}\psi'(\varphi)\eta \, d(t, x) \tag{5.40}$$

for all $\eta \in L^2(0, T; H^1(\Omega))$.

Proof We prove the theorem with the help of the so-called Faedo-Galerkin approximation which in essence means to project the equations and the solution spaces to a finite dimensional subspace of $H^1(\Omega)$. We now choose $(w_i)_{i\in\mathbb{N}}$ as eigenfunctions of the Laplace operator with Neumann boundary conditions. We refer to Jost [100, Section 11.5] for details on the eigenfunctions of elliptic operators. The functions w_i hence solve $-\Delta w_i = \lambda_i w_i$ in Ω, $\nabla w_i \cdot n = 0$ on $\partial\Omega$ in a weak sense, i.e.,

$$\int_{\Omega} \nabla w_i \cdot \nabla\zeta \, dx = \lambda_i \int_{\Omega} w_i \zeta \, dx \tag{5.41}$$

for all $\zeta \in H^1(\Omega)$. We normalize the w_i such that $(w_i, w_j)_{L^2(\Omega)} = \delta_{ij}$. As constant functions are eigenfunctions of the Laplace operator with eigenvalue 0, we can choose $\lambda_1 = 0$ and w_1 as the constant $\frac{1}{|\Omega|^{\frac{1}{2}}}$, where $|\Omega|$ is the Lebesgue measure of Ω. We now define the orthogonal projection

$$\pi_N : L^2(\Omega) \to X^N := span\{w_1, \dots, w_N\}$$

which is given as

$$\pi_N\varphi = \sum_{j=1}^{N} (\varphi, w_j)_{L^2(\Omega)} w_j.$$

It holds, see, e.g., [6],

$$\|\pi_N\varphi\|_{L^2(\Omega)} \leq \|\varphi\|_{L^2(\Omega)} \tag{5.42}$$

and

$$\|\nabla(\pi_N\varphi)\|_{L^2(\Omega)} \leq \|\nabla\varphi\|_{L^2(\Omega)}. \tag{5.43}$$

We now determine for all $t \in [0, T]$ functions $\varphi^N(t) \in \operatorname{span}\{w_1, \ldots, w_N\}$ and $\mu^N(t) \in \operatorname{span}\{w_1, \ldots, w_N\}$ as approximate solutions of (5.35), (5.36). In fact, we require

$$\varphi^N(t,x) = \sum_{i=1}^N c_i^N(t) w_i(x), \quad \mu^N(t,x) = \sum_{i=1}^N d_i^N(t) w_i(x)\,, \tag{5.44}$$

$$\int_\Omega \partial_t \varphi^N(t) w_j \, dx = -\int_\Omega \nabla \mu^N(t) \cdot \nabla w_j \, dx$$

$$\text{for } j = 1, \ldots, N\,, \tag{5.45}$$

$$\int_\Omega \mu^N(t) w_j \, dx = \int_\Omega \varepsilon \nabla \varphi^N(t) \cdot \nabla w_j \, dx + \int_\Omega \tfrac{1}{\varepsilon} \psi'(\varphi^N(t)) w_j \, dx$$

$$\text{for } j = 1, \ldots, N\,, \tag{5.46}$$

$$\varphi^N(0) = \pi_N \varphi_0. \tag{5.47}$$

Here, and in the whole book, we often write $v(t)(x)$ for a space-time function v, i.e., $v(t) : \Omega \to \mathbb{R}$. We use the orthonormality of the eigenfunctions $(w_i)_{i=1,\ldots,N}$ to compute

$$\int_\Omega \partial_t \varphi^N(t) w_j \, dx = \sum_{i=1}^N (c_i^N)'(t) \int_\Omega w_i w_j \, dx = (c_j^N)'(t)\,.$$

We obtain that the system (5.45), (5.46) reduces to a system of ordinary differential equations for the vector $\mathbf{c}^N(t) = (c_i^N(t))_{i=1,\ldots,N}$, where the right hand side of the ordinary differential equation depends continuously on $\mathbf{c}^N$. In fact, using (5.41) we obtain for $j = 1, \ldots, N$,

$$(c_j^N)' = -\lambda_j d_j^N\,,$$

$$d_j^N = \varepsilon \lambda_j c_j^N + \tfrac{1}{\varepsilon} \int_\Omega \psi'\left(\sum_{i=1}^N c_i^N w_i\right) w_j \, dx\,,$$

which has to hold for the vector valued function $\mathbf{c}^N$ together with the initial condition $c_j^N(0) = (\varphi_0, w_j)_{L^2(\Omega)}$. The second identity yields that a continuous function $F : \mathbb{R}^N \to \mathbb{R}^N$ exists such that $\mathbf{d}^N(t) = F(\mathbf{c}^N(t))$. Peano's theorem, see, e.g., Zeidler [154], now guarantees the existence of a local in time solution on a time interval $[0, \overline{T}]$ with a possibly small $\overline{T} > 0$.

In order to establish the existence on the whole time interval we show an energy estimate. For the free energy

$$\mathscr{F}(\varphi) = \int_\Omega \left(\tfrac{\varepsilon}{2} |\nabla \varphi|^2 + \tfrac{1}{\varepsilon} \psi(\varphi)\right) dx$$

we compute

$$\begin{aligned}
\frac{\mathrm{d}}{\mathrm{d}t}\mathscr{F}(\varphi^N(t)) &= \frac{\mathrm{d}}{\mathrm{d}t}\int_\Omega \left(\tfrac{\varepsilon}{2}|\nabla\varphi^N(t)|^2 + \tfrac{1}{\varepsilon}\psi(\varphi^N(t))\right)dx \\
&= \int_\Omega (\varepsilon\nabla\varphi^N(t)\cdot\nabla\partial_t\varphi^N(t) + \tfrac{1}{\varepsilon}\psi'(\varphi^N(t))\partial_t\varphi^N(t))dx \\
&= \int_\Omega \mu^N(t)\partial_t\varphi^N(t)dx \\
&= -\int_\Omega |\nabla\mu^N(t)|^2 dx\,.
\end{aligned}$$

In the second equality we used the fact that $c_1^N, \ldots, c_N^N$ are C^1-functions with respect to t. The third equality follows if we multiply (5.46) by $(c_j^N)'(t)$ and sum over all j. The last identity follows by multiplying (5.45) by $d_j^N(t)$ and summing over j.

Integrating the above equality with respect to t gives for all $s \in [0, \overline{T}]$ and all $N \in \mathbb{N}$

$$\begin{aligned}
&\int_\Omega \tfrac{\varepsilon}{2}|\nabla\varphi^N(s)|^2dx + \int_{\Omega_s}|\nabla\mu^N|^2 d(t,x) \\
&= \int_\Omega \tfrac{\varepsilon}{2}\left(|\nabla\varphi^N(0)|^2 + \frac{1}{\varepsilon}\psi(\varphi^N(0))\right)dx - \int_\Omega \tfrac{1}{\varepsilon}\psi(\varphi^N(s))dx \le C\,. \qquad (5.48)
\end{aligned}$$

We need to verify the last inequality which states that a C not depending on N exists such that this inequality is true. As $\varphi_0 \in H^1(\Omega)$ we obtain from (5.42) and (5.43) that $\pi_N\varphi_0 = \varphi^N(0)$ is uniformly bounded in $H^1(\Omega)$. By the Sobolev embedding theorem the Sobolev space $H^1(\Omega)$ embeds into $L^p(\Omega)$ with $p = \frac{2n}{n-2}$ for $n \ge 3$ and p arbitrary for $n = 1, 2$. From Assumption (A), the fundamental theorem of calculus and Young's inequality we can now derive that suitable constants C_4 and C_5 exist such that for all $z \in \mathbb{R}$

$$|\psi(z)| \le C_4|z|^q + C_5\,,$$

where $q < \frac{2n}{n-2}$ if $n \ge 3$ and $q \in \mathbb{R}^+$ arbitrary if $n = 1, 2$. Combining the above arguments we hence obtain that $\int_\Omega \psi(\varphi^N(0))dx$ is uniformly bounded. This together with the fact that $\psi \ge -C_3$ then implies that the constant in (5.48) can be chosen independent of N.

As w_1 is constant we obtain from Eq. (5.45) for $j = 1$ that

$$\frac{\mathrm{d}}{\mathrm{d}t}\int_\Omega \varphi^N(t)dx = 0\,. \qquad (5.49)$$

In addition, noting that for $i \geq 2$ the orthogonality $(w_i, w_1)_{L^2} = 0$ implies $\int_\Omega w_i dx = 0$, we have

$$\int_\Omega \varphi^N(0)dx = \sum_{i=1}^{N} (\varphi_0, w_i)_{L^2(\Omega)} \int_\Omega w_i\, dx = (\varphi_0, w_1)_{L^2(\Omega)} \int_\Omega w_1\, dx$$

$$= (\varphi_0, |\Omega|^{-\frac{1}{2}})_{L^2(\Omega)} |\Omega|^{-\frac{1}{2}} |\Omega| = \int_\Omega \varphi_0\, dx .$$

Hence, we obtain $\int_\Omega \varphi^N(0)dx = \int_\Omega \varphi_0\, dx$, which does not depend on N. This together with (5.49) yields

$$\int_\Omega \varphi^N(t)dx = \int_\Omega \varphi_0\, dx .$$

From estimate (5.48) we can now deduce that $\nabla\varphi^N \in L^\infty(0, T; L^2(\Omega))$ and Poincaré inequality for functions with mean value zero then yields

$$\sup_{t\in[0,\overline{T}]} \|\varphi^N(t)\|_{H^1(\Omega)} \leq C .$$

As all the $\varphi^N(t)$ lie in a finite dimensional vector space on which all norms are equivalent, we obtain that the $(c_1^N, \ldots, c_N^N)$ are bounded in $[0, \overline{T}]$. Now ODE theory implies that the local solution can be extended to the full interval $[0, T]$.

Denoting, as above, by π_N the orthogonal projection of $L^2(\Omega)$ onto $\operatorname{span}\{w_1, \ldots, w_N\}$, we obtain for all $\zeta \in L^2(0, T; H^1(\Omega))$

$$\begin{aligned}
\left| \int_{\Omega_T} \partial_t \varphi^N \zeta\, d(t, x) \right| &= \left| \int_{\Omega_T} \partial_t \varphi^N \pi_N \zeta\, d(t, x) \right| \\
&= \left| \int_{\Omega_T} \nabla\mu^N \cdot \nabla(\pi_N \zeta) d(t, x) \right| \\
&\leq \left(\int_{\Omega_T} |\nabla\mu^N|^2 d(t, x) \right)^{\frac{1}{2}} \left(\int_{\Omega_T} |\nabla(\pi_N \zeta)|^2 d(t, x) \right)^{\frac{1}{2}} \\
&\leq \left(\int_{\Omega_T} |\nabla\mu^N|^2 d(t, x) \right)^{\frac{1}{2}} \|\nabla\zeta\|_{L^2(\Omega_T)} \\
&\leq C \|\nabla\zeta\|_{L^2(\Omega_T)} ,
\end{aligned} \tag{5.50}$$

where we compute for all t the L^2-projection $\pi_N\zeta$ with respect to the x-variable and use (5.43) in the second inequality. This implies that $\partial_t\varphi^N$ is uniformly bounded in $L^2(0,T;(H^1(\Omega))')$, i.e., there exists a $C>0$, such that

$$\|\partial_t\varphi^N\|_{L^2(0,T;(H^1(\Omega))')}\le C\,.$$

We now need to use a compactness result by Aubin et al., see [29, 142]. It is given as follows. Let X, Y and Z be Banach spaces with a compact embedding $X \hookrightarrow Y$ and a continuous embedding $Y \hookrightarrow Z$. Then the embeddings

$$\{u\in L^2(0,T;X)\mid \partial_t u\in L^2(0,T;Z)\}\hookrightarrow L^2(0,T;Y) \tag{5.51}$$

and

$$\{u\in L^\infty(0,T;X)\mid \partial_t u\in L^2(0,T;Z)\}\hookrightarrow C^0([0,T];Y) \tag{5.52}$$

are compact. We now apply this result for the case $X=H^1(\Omega)$, $Y=L^2(\Omega)$ ($Y=L^p(\Omega)$ with $2\le p<\frac{2n}{n-2}$, respectively), and $Z=(H^1(\Omega))'$. With this result we obtain for suitable subsequences

$$\begin{aligned}
\varphi^N &\to \varphi && \text{weakly} && \text{in } L^2(0,T;H^1(\Omega))\,,\\
\varphi^N &\to \varphi && \text{strongly} && \text{in } C^0([0,T];L^2(\Omega))\,,\\
\partial_t\varphi^N &\to \partial_t\varphi && \text{weakly} && \text{in } L^2(0,T;(H^1(\Omega))') \text{ and}\\
\varphi^N &\to \varphi && \text{strongly} && \text{in } L^2(0,T;L^p(\Omega)) \quad \text{and a.e. in } \Omega_T\,,
\end{aligned}$$

where $p<\frac{2n}{n-2}$. We can now use Assumption (A), the almost everywhere convergence of φ^N, the fact that φ^N converges strongly in $L^2(0,T;L^p(\Omega))$, for all p as above, and Lebesgue's general convergence theorem, see [6], to deduce that $\psi'(\varphi^N)\to\psi'(\varphi)$ in $L^2(\Omega_T)$.

It remains to show the convergence of μ^N. Choosing $j=1$ in (5.46) gives $\int_\Omega \mu^N(t)=\int_\Omega\frac{1}{\varepsilon}\psi'(\varphi^N(t))$. Now Assumption (A), the fact that φ^N is uniformly bounded in $L^\infty(0,T;L^p(\Omega))$ with $p=\frac{2n}{n-2}$ and Poincaré's inequality for functions with mean value zero gives

$$\|\mu^N\|_{L^2(0,T;H^1(\Omega))}\le C\,.$$

This implies (again for a subsequence)

$$\mu^N\to\mu \quad \text{weakly in} \quad L^2(0,T;H^1(\Omega))\,.$$

It is possible to deduce from (5.45), (5.46) that

$$\int_0^T \langle \partial_t \varphi^N(t), \zeta(t) \rangle_{(H^1)', H^1} dt = -\int_{\Omega_T} \nabla \mu^N \cdot \nabla \zeta \, d(t,x) \tag{5.53}$$

for all $\zeta \in L^2(0, T; \text{span}\{w_1, \ldots, w_N\})$ and

$$\int_{\Omega_T} \mu^N \eta \, d(t,x) = \int_{\Omega_T} \varepsilon \nabla \varphi^N \cdot \nabla \eta \, d(t,x) + \int_{\Omega_T} \tfrac{1}{\varepsilon} \psi'(\varphi^N) \eta \, d(t,x) \tag{5.54}$$

for all $\eta \in L^2(0, T; \text{span}\{w_1, \ldots, w_N\})$. Using the convergence properties of (φ^N, μ^N) we can now pass to the limit in (5.53) and (5.54) to obtain (5.39) and (5.40) for all $\zeta \in L^2(0, T; \text{span}\{w_i \mid i \in \mathbb{N}\})$ and $\eta \in L^2(0, T; \text{span}\{w_i \mid i \in \mathbb{N}\})$. As $\text{span}\{w_i \mid i \in \mathbb{N}\}$ is dense in $H^1(\Omega)$ we also obtain (5.53)–(5.54) for all $\zeta, \eta \in L^2(0, T; H^1(\Omega))$. The latter follows since functions of the form $\sum_{i=1}^N \alpha_i(t) w_i(x)$, with $N \in \mathbb{N}$ and $\alpha_i \in L^2(0, T)$, $i = 1, \ldots N$, are dense in $L^2(0, T; H^1(\Omega))$. The strong convergence of φ^N in $C^0([0, T]; L^2(\Omega))$ and the fact that $\varphi^N(0) \to \varphi_0$ in $L^2(\Omega)$ gives $\varphi(0) = \varphi_0$. This proves the theorem. □

5.9 The Mullins–Sekerka Problem as the Sharp Interface Limit of the Cahn–Hilliard Equation

In this section we identify the sharp interface limit of the Cahn–Hilliard equation. In fact, we will obtain the Mullins–Sekerka problem (5.17)–(5.20) in the limit, when the thickness of the interface (which is proportional to ε) in the Cahn-Hilliard model tends to zero. To do so, we will use the method of formally matched asymptotic expansions, where asymptotic expansions in bulk regions have to match with expansions in interfacial regions.

5.9.1 The Governing Equations

As usual for phase field models, we introduce a scaling for the free energy with respect to a small length scale parameter ε as follows

$$\frac{\hat{\gamma}\varepsilon}{2} |\nabla \varphi|^2 + \frac{\hat{\gamma}}{\varepsilon} \psi(\varphi),$$

where $\hat{\gamma}$ is a constant related to the surface energy density γ. We now consider the following Cahn–Hilliard system:

$$\partial_t \varphi = \nabla \cdot (m_0 \nabla \mu) , \tag{5.55}$$

$$\mu = \frac{\hat{\gamma}}{\varepsilon} \psi'(\varphi) - \hat{\gamma} \varepsilon \Delta \varphi . \tag{5.56}$$

We assume that

- $m_0 > 0$ is constant,
- $\psi(\varphi)$ is a double-well potential such that $\psi(1) = \psi(-1) = 0$ and $\psi(z) > 0$ if $z \notin \{1, -1\}$.

For a solution $(\varphi^\varepsilon, \mu^\varepsilon)$ of the system (5.55) and (5.56) we perform formally matched asymptotic expansions. It will turn out that the phase field φ^ε will change its values rapidly on a length scale proportional to ε. For additional information on asymptotic expansions for phase field equations we refer to [71, 78].

5.9.2 Outer Expansions

In a first step we expand the solution in outer regions away from the interface. We assume expansions of the form $\varphi^\varepsilon = \sum_{k=0}^\infty \varepsilon^k \varphi_k$ and $\mu^\varepsilon = \sum_{k=0}^\infty \varepsilon^k \mu_k$. We now plug these expansions in Eqs. (5.55) and (5.56) and obtain

$$\partial_t \left(\sum_{k=0}^\infty \varepsilon^k \varphi_k \right) = \nabla \cdot \left(m_0 \nabla \left(\sum_{k=0}^\infty \varepsilon^k \mu_k \right) \right) , \tag{5.57}$$

$$\sum_{k=0}^\infty \varepsilon^k \mu_k = \frac{\hat{\gamma}}{\varepsilon} \psi' \left(\sum_{k=0}^\infty \varepsilon^k \varphi_k \right) - \hat{\gamma} \varepsilon \Delta \left(\sum_{k=0}^\infty \varepsilon^k \varphi_k \right) \tag{5.58}$$

and solve the equations order by order. We first expand Eq. (5.58) in outer regions. Here, a Taylor expansion of the ψ'-term gives

$$\frac{\hat{\gamma}}{\varepsilon} \psi' \left(\sum_{k=0}^\infty \varepsilon^k \varphi_k \right) = \frac{\hat{\gamma}}{\varepsilon} \left(\psi'(\varphi_0) + \psi''(\varphi_0) \left(\sum_{k=1}^\infty \varepsilon^k \varphi_k \right) + \dots \right)$$

and hence (5.58) gives to leading order, i.e., after setting all terms multiplying $\frac{1}{\varepsilon}$ to zero,

$$\psi'(\varphi_0) = 0 .$$

As stable solutions of this equation we obtain ± 1. The fact that we obtain ± 1 can also be seen from the fact that

$$\frac{\mathrm{d}}{\mathrm{d}t}\int_\Omega \left(\tfrac{\varepsilon}{2}|\nabla\varphi|^2 + \tfrac{1}{\varepsilon}\psi(\varphi)\right) dx \le 0\,, \tag{5.59}$$

i.e., $\int_\Omega \frac{1}{\varepsilon}\psi(\varphi)dx$ is bounded for $t > 0$ if the energy $\int_\Omega \left(\frac{\varepsilon}{2}|\nabla\varphi|^2 + \frac{1}{\varepsilon}\psi(\varphi)\right) dx$ is bounded initially. The fact that $\psi(z)$ is only 0 if $z = \pm 1$ then implies that $\varphi = \pm 1$ needs to hold for $\varepsilon \to 0$ almost everywhere.

At each time t we will denote by $\Omega^{\pm}(t)$ the regions where $\varphi_0 = \pm 1$. Using that $\varphi_0 = \pm 1$ we obtain from (5.57) to leading order ε^0:

$$\Delta\mu_0 = 0 \quad \text{in} \qquad \Omega^{\pm}(t),$$

i.e., in this case we set all terms multiplying ε^0 to zero and use that $\partial_t\varphi_0 = 0$.

5.9.3 Inner Expansions

In the next step we make an expansion in an interfacial region, where a transition between two phases takes place.

New Coordinates in the Inner Region

We denote by $\Gamma = (\Gamma(t))_{t\in[0,T]}$ the smoothly evolving interface, which we expect to be the limit of the zero level sets of φ when ε tends to zero and will now introduce new coordinates in a neighborhood of Γ. Choosing the time interval $I \subset [0, T]$ and a spatial parameter domain $U \subset \mathbb{R}^{n-1}$ we define a local parametrization

$$F : I \times U \to \mathbb{R}^n$$

of Γ. By ν we denote the unit normal to $\Gamma(t)$ pointing into phase 2 (which is the phase related to $\varphi = 1$). Close to $F(I \times U)$ we consider the signed distance function $d_\Gamma(t, x)$ of a point x to $\Gamma(t)$ with $d_\Gamma(t, x) > 0$ if $x \in \Omega^+(t)$. Similar as in Sect. 2.4 we now introduce a local parametrization of $I \times \mathbb{R}^n$ close to $F(I \times U)$ using the rescaled distance $z = \frac{d_\Gamma}{\varepsilon}$ as follows

$$G^\varepsilon(t, s, z) := (t, F(t, s) + \varepsilon z\nu(t, s)),$$

where $s \in U \subset \mathbb{R}^{n-1}$. We denote by

$$V = \partial_t F \cdot \nu$$

the normal velocity and observe that the inverse function $(t, s, z)(t, x) := (G^\varepsilon)^{-1}(t, x)$ fulfills

$$\partial_t z = \tfrac{1}{\varepsilon}\partial_t d_\Gamma = -\tfrac{1}{\varepsilon}V\,.$$

To derive the last identity we used (5.2) and the fact that $|\nabla d_\Gamma| = 1$. For a scalar function $b(t, x)$ we obtain for $\hat{b}$ defined in the new coordinates via $\hat{b}(t, s(t, x), z(t, x)) = b(t, x)$ the identity

$$\frac{d}{dt}b(t, x) = \partial_t z \partial_z \hat{b} + \partial_t s \cdot \nabla_s \hat{b} + \partial_t \hat{b} = -\tfrac{1}{\varepsilon}V\partial_z \hat{b} + \text{ h.o.t.}, \tag{5.60}$$

where h.o.t. stands for terms that are higher order in ε. With respect to the spatial variables we obtain, see the Appendix of [3],

$$\nabla_x b = \nabla_{\Gamma_{\varepsilon z}} \hat{b} + \tfrac{1}{\varepsilon}\partial_z \hat{b}\,\nu \tag{5.61}$$

with $\nabla_{\Gamma_{\varepsilon z}}$ the surface gradient on

$$\Gamma_{\varepsilon z} := \{F(s) + \varepsilon z \nu(s) \mid s \in U\},$$

where here and in what follows we often omit the t-dependence. For a vector quantity $j(t, x)$ written in the new coordinates via $\hat{j}(t, s(t, x), z(t, x)) = j(t, x)$ we obtain

$$\nabla_x \cdot j = \nabla_{\Gamma_{\varepsilon z}} \cdot \hat{j} + \tfrac{1}{\varepsilon}\partial_z \hat{j} \cdot \nu, \tag{5.62}$$

where $\nabla_{\Gamma_{\varepsilon z}} \cdot \hat{j}$ is the divergence of $\hat{j}$ on $\Gamma_{\varepsilon z}$. In the Appendix of [3] it was computed that

$$\Delta_x b = \Delta_{\Gamma_{\varepsilon z}} \hat{b} - \tfrac{1}{\varepsilon}(\kappa + \varepsilon z|H|^2)\partial_z \hat{b} + \tfrac{1}{\varepsilon^2}\partial_{zz}\hat{b} + \text{ h.o.t.}\,, \tag{5.63}$$

where κ is the mean curvature (the sum of the principal curvatures) and $|H|$ is the Frobenius norm of the Weingarten map H. In addition, we note that (see the Appendix of [3])

$$\nabla_{\Gamma_{\varepsilon z}} \hat{b}(s, z) = \nabla_\Gamma \hat{b}(s, z) + \text{ h.o.t.}\,, \tag{5.64}$$

$$\nabla_{\Gamma_{\varepsilon z}} \cdot \hat{j}(s, z) = \nabla_\Gamma \cdot \hat{j}(s, z) + \text{ h.o.t.}\,, \tag{5.65}$$

$$\Delta_{\Gamma_{\varepsilon z}} \hat{b}(s, z) = \Delta_\Gamma \hat{b}(s, z) + \text{ h.o.t.}\,, \tag{5.66}$$

where ∇_Γ, $\nabla_\Gamma \cdot$, Δ_Γ are the surface gradient, the surface divergence and the surface Laplacian on Γ, respectively.

Matching Conditions

We now assume an ε-series approximation of the unknown functions φ, μ, which in the inner variables we will denote by Φ, M. Denoting by $\Phi_0 + \varepsilon\Phi_1 + \ldots$ the inner expansion and by $\varphi_0 + \varepsilon\varphi_1 + \ldots$ the outer expansion of the phase field, we obtain the following matching conditions at $x = F(s)$:

$$\lim_{z\to\pm\infty} \Phi_0(z,s) = \varphi_0(x\pm)\,, \tag{5.67}$$

$$\lim_{z\to\pm\infty} \partial_z\Phi_1(z,s) = \nabla\varphi_0(x\pm)\cdot\nu, \tag{5.68}$$

where $\varphi_0(x\pm), \ldots$ denotes the limit $\lim\limits_{\delta\searrow 0}\varphi_0(x \pm \delta\nu)$. In addition, we obtain that if $\Phi_1(z,s) = A_\pm(s) + B_\pm(s)z + o(1)$ as $z \to \pm\infty$ the identities

$$A_\pm(s) = \varphi_1(x\pm),\;\; B_\pm(s) = \nabla\varphi_0(x\pm)\cdot\nu \tag{5.69}$$

have to hold (see [68, 77]). Of course, similar relations hold for the function μ.

The Equations to Leading Order

Plugging the asymptotic expansions into (5.55) and (5.56), we ask that each individual coefficient of a power in ε vanishes. Equation (5.56) gives to leading order $\frac{1}{\varepsilon}$, where we again use a Taylor expansion for the ψ'-term and also use (5.63) which expresses the Laplace operator in the inner variables,

$$0 = \partial_{zz}\Phi_0 - \psi'(\Phi_0)\,. \tag{5.70}$$

From (5.67) and the fact that $\varphi_0 = \pm 1$ in the outer regions we obtain that the following limits are attained:

$$\Phi_0(z) \to \pm 1 \quad \text{for} \quad z \to \pm\infty\,. \tag{5.71}$$

Here, the inner variable Φ_0 has to match the outer variable φ_0. We now choose the unique solution of (5.70), (5.71) which fulfills

$$\Phi_0(0) = 0\,.$$

We in particular obtain that Φ_0 does not depend on t and s.

Equation (5.55) now gives to leading order $\frac{1}{\varepsilon^2}$

$$0 = m_0\partial_{zz}M_0\,.$$

Matching implies that M_0 needs to converge to the outer solution for $z \to \pm\infty$. In particular, we obtain that Φ_0 is bounded and hence, due to the fact that M_0 is also affine linear, we obtain that M_0 is constant. In addition, we derive, after again matching with the outer solution, that

$$[\mu_0]_-^+ = 0\,.$$

The Equation for the Chemical Potential at the Interface

The equation for the chemical potential gives to the order ε^0

$$M_0 = \hat{\gamma}\psi''(\Phi_0)\Phi_1 - \hat{\gamma}\partial_{zz}\Phi_1 + \hat{\gamma}\partial_z\Phi_0\kappa\,. \tag{5.72}$$

In order to be able to obtain a solution Φ_1 from (5.72), a solvability condition has to hold. This solvability condition will yield the generalized Gibbs–Thomson equation. To see this, we multiply (5.72) with $\partial_z\Phi_0$, integrate with respect to z and obtain (using the facts that M_0 and V_0 do not depend on z):

$$2M_0 = \hat{\gamma}\int_{-\infty}^{\infty}(\psi''(\Phi_0)\partial_z\Phi_0\Phi_1 - \partial_{zz}\Phi_1\partial_z\Phi_0)dz + \hat{\gamma}\kappa\int_{-\infty}^{\infty}(\partial_z\Phi_0)^2dz\,.$$

Defining

$$c_0 := \int_{-\infty}^{\infty}(\partial_z\Phi_0)^2\,dz,$$

we obtain after integration by parts, using the fact that $\partial_z\Phi_0(z)$, $\partial_{zz}\Phi_0(z)$ decay exponentially for $|z| \to \infty$,

$$2M_0 = \hat{\gamma}\int_{-\infty}^{\infty}\partial_z(\psi'(\Phi_0) - \partial_{zz}\Phi_0)\Phi_1dz + \hat{\gamma}c_0\kappa.$$

Since $\psi'(\Phi_0) - \partial_{zz}\Phi_0 = 0$, see (5.70), we obtain after matching

$$2\mu_0 = \gamma\kappa,$$

where $\gamma := c_0\hat{\gamma}$.

Interfacial Flux Balance in the Sharp Interface Limit

We now expand Eq. (5.55) further in order to obtain contributions of the diffusive fluxes at the interface.

At order $\frac{1}{\varepsilon}$ we deduce from (5.55), where we used (5.60) and (5.66),

$$(-V)\partial_z\Phi_0 = \partial_z(m_0\partial_z M_1)\,. \tag{5.73}$$

Matching, see (5.68), gives $\partial_z M_1 \to \nabla \mu_0 \cdot \nu$ for $z \to \pm\infty$. Integrating (5.73) gives

$$-2V = m_0[\nabla \mu_0 \cdot \nu]_-^+ . \tag{5.74}$$

Altogether, we now obtained the following version of the Mullins–Sekerka problem (dropping the index 0):

$$-\Delta\mu = 0 \quad \text{in } \Omega_-(t) \cup \Omega_+(t)\,, \tag{5.75}$$

$$2V = -m_0[\nabla\mu]_-^+ \cdot \nu \quad \text{on } \Gamma(t)\,, \tag{5.76}$$

$$2\mu = \gamma\kappa \quad \text{on } \Gamma(t)\,, \tag{5.77}$$

$$\nabla\mu \cdot n = 0 \quad \text{on } \partial\Omega\,. \tag{5.78}$$

5.10 How to Discretize the Cahn–Hilliard Equation?

In this section we would like to demonstrate that phase field approaches like the Cahn–Hilliard equation can easily be solved numerically by using standard finite element approaches and suitable time discretizations. We refer to [13,20,50,63,79] for more details.

5.10.1 The Time Discrete Setting

We introduce a time discretization which mimics the energy inequality for the Cahn–Hilliard equation and conserves the total integral, see (3.63) and (5.38), on the discrete level. Let $0 = t_0 < t_1 < \ldots < t_{k-1} < t_k < t_{k+1} < \ldots < t_M = T$ denote an equidistant subdivision of the interval $\overline{I} = [0, T]$ with $t_{k+1} - t_k = \delta = T/M$. From here onwards the superscript k denotes the corresponding variables at time instance t_k.

Time Integration Scheme

Let $\varphi_0 \in H^1(\Omega)$ and set $\varphi^0 = \varphi_0$.

Time discrete systems for k ≥ 1:

Given $\varphi^k \in H^1(\Omega)$,

find $\varphi^{k+1} \in H^1(\Omega)$, $\mu^{k+1} \in H^1(\Omega)$ satisfying

$$\frac{1}{\delta}\int_\Omega (\varphi^{k+1} - \varphi^k)\eta\, dx + \int_\Omega m(\varphi^k)\nabla\mu^{k+1} \cdot \nabla\eta\, dx = 0 \qquad \forall \eta \in H^1(\Omega), \tag{5.79}$$

$$\begin{aligned} &\epsilon \int_\Omega \nabla\varphi^{k+1} \cdot \nabla\zeta\, dx - \int_\Omega \mu^{k+1}\zeta\, dx \\ &+ \frac{1}{\epsilon}\int_\Omega ((\psi_+)'(\varphi^{k+1}) + (\psi_-)'(\varphi^k))\zeta\, dx = 0 \qquad \forall \zeta \in H^1(\Omega), \end{aligned} \tag{5.80}$$

where ψ_+ is convex and ψ_- is concave such that $\psi = \psi_+ + \psi_-$. We remark that it is always possible to find such a so-called convex-concave splitting.

Remark 5.10.1

(i) We note that in (5.79) and (5.80) the only nonlinearity arises from Ψ'_+ and thus only Eq. (5.80) is nonlinear.
(ii) In this section we always assume that the mobility m is continuous and bounded from below by $\underline{m} > 0$.

5.10.2 The Fully Discrete Setting

For a numerical treatment we next discretize the weak formulation (5.79) and (5.80) in space.

Let $\mathscr{T} = (T_i)_{i=1}^{N_T}$ denote a conforming triangulation of $\overline{\Omega}$ with closed simplices $T_i, i = 1, \ldots, N_T$, edges $E_i, i = 1, \ldots, N_E$, $\mathscr{E} = \bigcup_{i=1}^{N_E} E_i$. We also denote by N_P the number of vertices and by $\Phi^i, i = 1, \ldots N_P$, the standard piecewise linear Lagrange basis functions. On $\mathscr{T}$ we define the following finite element space:

$$X_h = \{v \in C(\overline{\Omega}) \mid v|_T \in P_1(T)\, \forall T \in \mathscr{T}\} =: \operatorname{span}\{\Phi^i\}_{i=1}^{N_P},$$

where $P_l(S)$ denotes the space of polynomials up to order $l \in \mathbb{N}$ defined on S. Above h refers to the maximal diameter of the simplices $T_i, i = 1, \ldots, N_T$.

Using this finite element space we state the discrete counterpart of (5.79) and (5.80):

Let $k \geq 1$, given $\varphi_h^k \in X_h$, find $\varphi_h^{k+1} \in X_h$, $\mu_h^{k+1} \in X_h$ such that for all $\zeta \in X_h$, $\eta \in X_h$ there holds:

$$\frac{1}{\delta}(\varphi_h^{k+1} - \varphi_h^k, \zeta) + (m(\varphi_h^k)\nabla\mu_h^{k+1}, \nabla\zeta) = 0, \tag{5.81}$$

$$\epsilon(\nabla\varphi_h^{k+1}, \nabla\eta) + \frac{1}{\epsilon}(\Psi'_+(\varphi_h^{k+1}) + \Psi'_-(\varphi_h^k), \eta) - (\mu_h^{k+1}, \eta) = 0, \tag{5.82}$$

where $\varphi_h^0 = P\varphi_0$ denotes the L^2 projection of φ_0 in X_h and $(\cdot, \cdot)$ is the L^2-inner product. Another possibility is to choose φ_h^0 as the Lagrange interpolation of φ_0.

Remark 5.10.2

(i) The system (5.81) and (5.82) is a finite dimensional nonlinear system in which $\Psi'_+(\varphi_h^{k+1})$ is the only non-linear term. The complete system can be solved using Newton's method. For the resulting linear systems in each time step sparse direct solvers, see for instance [41], can be used provided that the number of unknowns is not too large, lets say less than 500,000 degrees of freedom. For systems with more unknowns one needs iterative methods with preconditioning. In this case one can for example use algebraic multigrid (AMG) preconditioners to accelerate the solution of the linear systems, see [28], and the literature therein for details.

(ii) We remark that the system (5.81) and (5.82) is much easier to be solved when compared to methods based on parametric methods and also allows for topology changes.

(iii) In Fig. 5.6 we plot a numerical solution of the Cahn–Hilliard equation with plenty of topology changes.

(iv) It is also possible to consider phase field equations based on an anisotropic Ginzburg–Landau energy, see [20, 85]. We refer to Fig. 5.7 for numerical computations in the anisotropic case.

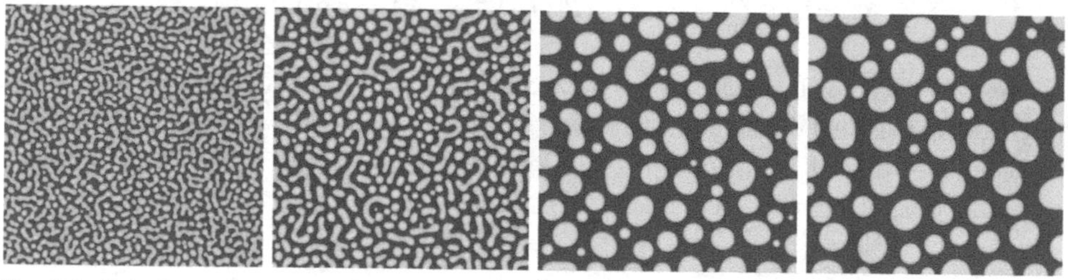

Fig. 5.6 Solutions of the Cahn–Hilliard equation with Neumann boundary conditions for φ and $\Delta\varphi$. Computations by Dennis Trautwein

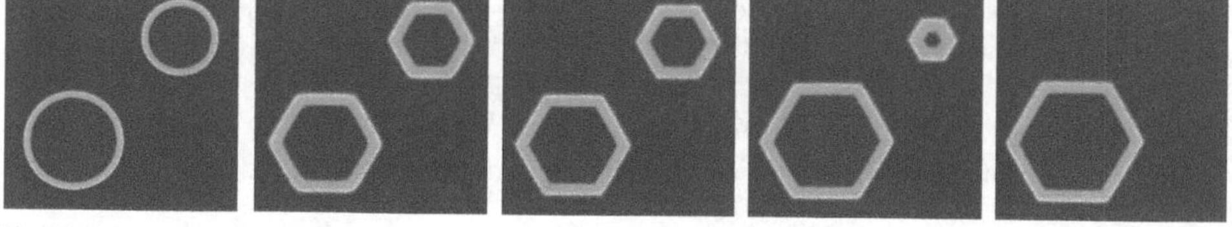

Fig. 5.7 A solution of the anisotropic Cahn–Hilliard equation in two space dimensions with an anisotropy with hexagonal symmetry, cf. [20] for details

5.10.3 Existence of Solutions to the Fully Discrete System

We next show the existence of a solution to the fully discrete system (5.81) and (5.82).

Theorem 5.10.3 *There exist $\varphi_h^{k+1} \in X_h$, $\mu_h^{k+1} \in X_h$ solving* (5.81) *and* (5.82).

Proof Any solution of the system (5.81) and (5.82) fulfills

$$(\varphi_h^{k+1}, 1) = (\varphi_h^k, 1)$$

which one obtains by testing Eq. (5.81) with $\zeta \equiv 1$. Therefore, the mean value of φ_h^{k+1} is fixed and we only need to specify the mean value free part of φ_h^{k+1}. We hence define $\alpha = \frac{1}{|\Omega|}\int_\Omega \varphi_h^k \, dx$ and set

$$V_{(0)} := \{v_h \in X_h \mid (v_h, 1) = 0\}.$$

Then $z^{k+1} := \varphi_h^{k+1} - \alpha$ fulfills $z^{k+1} \in V_{(0)}$. In the following we use z^{k+1} as unknown for the phase field, since the mean value of φ is fixed. In addition, we introduce $y^{k+1} := \mu_h^{k+1} - \frac{1}{|\Omega|}\int_\Omega \mu_h^{k+1} \, dx$ and require (5.81) and (5.82) preliminarily only for test functions with zero mean value.

We define

$$X = V_{(0)} \times V_{(0)}$$

with the inner product

$$((y_1, z_1), (y_2, z_2))_X := (\nabla y_1, \nabla y_2) + (\nabla z_1, \nabla z_2)$$

and norm $\|\cdot\|_X^2 = (\cdot, \cdot)_X$. It follows from the inequality of Poincaré for functions with mean value zero that $(\cdot, \cdot)_X$ indeed defines an inner product on X. For $(y, z) \in X$ we define

$$\begin{aligned}(G(y, z), (\overline{y}, \overline{z}))_X &:= (z - \varphi_h^k, \overline{y}) + \delta(m(\varphi_h^k)\nabla y, \nabla \overline{y}) + \epsilon(\nabla z, \nabla \overline{z}) \\ &\quad + \frac{1}{\epsilon}(\Psi'_+(z + \alpha) + \Psi'_-(\varphi_h^k), \overline{z}) - (y, \overline{z}). \qquad (5.83)\end{aligned}$$

The function G is obviously continuous. Now we show $(G(y, z), (y, z))_X > 0$ for $\|(y, z)\|_X$ large enough. It will then follow from [147, Lem. II.1.4], which is a consequence of Brouwer's fixed point theorem and will be stated in the following remark, that G admits a root $(y^*, z^*) \in X$.

We estimate

$$(G(y,z),(y,z))_X \geq \delta \underline{m}(\nabla y, \nabla y) + \epsilon(\nabla z, \nabla z) + \frac{1}{\epsilon}(\Psi'_+(z+\alpha), z) \\ -(\varphi_h^k, y) + \frac{1}{\epsilon}(\Psi'_-(\varphi_h^k), z). \tag{5.84}$$

Using the convexity of Ψ_+, which implies that Ψ'_+ is monotone, one obtains

$$(\Psi'_+(z+\alpha), z) = (\Psi'_+(z+\alpha) - \Psi'_+(\alpha), z) + (\Psi'_+(\alpha), z) \geq (\Psi'_+(\alpha), z).$$

By using Young's and Poincaré's inequality in (5.84) we obtain

$$(G(y,z),(y,z))_X > 0$$

for $\|(y,z)\|_X \geq R$, if R is large enough. Now [147, Lem. II.1.4] implies the existence of $(y^*, z^*) \in X$ such that $G(y^*, z^*) = 0$. Setting $(\mu, \varphi) = (y^* + \beta, z^* + \alpha)$ with β such that $(\beta, 1) = \frac{1}{\epsilon}(\Psi'_+(\varphi) + \Psi'_-(\varphi_h^k), 1)$ is then a solution of (5.81) and (5.82). □

Remark 5.10.4

(i) Note that we do not need any smallness requirement on the mesh size h or on the time step length δ.

(ii) Lemma II.1.4 of [147] states:
Let X be a finite dimensional Hilbert space with inner product $\langle .,. \rangle$ and corresponding norm $\|.\|$ and let F be a continuous map from X into itself such that

$$\langle F(x), x \rangle > 0 \quad \text{for} \quad \|x\| = R > 0$$

for some $R > 0$. Then there exists an $x^* \in X$, $\|x^*\| \leq R$, such that

$$F(x^*) = 0.$$

5.10.4 An Energy Inequality in the Fully Discrete Setting

We will now show that the fully discrete system fulfills an energy inequality similar as in the continuous case. This leads to estimates which together with compactness theorems can be used to show that the discrete solutions converge to a continuous solution.

Theorem 5.10.5 *Let $(\varphi_h^{k+1}, \mu_h^{k+1})$ be a solution to (5.81) and (5.82). Then it holds for all $k \geq 1$:*

$$\begin{aligned}
&\frac{\epsilon}{2}\int_\Omega |\nabla\varphi_h^{k+1}|^2\,dx + \frac{1}{\epsilon}\int_\Omega \Psi(\varphi_h^{k+1})\,dx \\
&\qquad + \frac{\epsilon}{2}\int_\Omega |\nabla\varphi_h^{k+1} - \nabla\varphi_h^{k}|^2\,dx + \delta\int_\Omega m(\varphi_h^k)|\nabla\mu_h^{k+1}|^2\,dx \\
&\qquad \leq \frac{\epsilon}{2}\int_\Omega |\nabla\varphi_h^{k}|^2\,dx + \frac{1}{\epsilon}\int_\Omega \Psi(\varphi_h^{k})\,dx. \qquad (5.85)
\end{aligned}$$

Proof We have

$$\begin{aligned}
&\nabla\varphi_h^{k+1}\cdot\left(\nabla\varphi_h^{k+1} - \nabla\varphi_h^{k}\right) \\
&\qquad = \frac{1}{2}|\nabla\varphi_h^{k+1}|^2 - \frac{1}{2}|\nabla\varphi_h^{k}|^2 + \frac{1}{2}|\nabla\varphi_h^{k+1} - \nabla\varphi_h^{k}|^2, \qquad (5.86)
\end{aligned}$$

and since Ψ_+ is convex and Ψ_- is concave,

$$\Psi_+(\varphi_h^{k+1}) - \Psi_+(\varphi_h^{k}) \leq \Psi_+'(\varphi_h^{k+1})(\varphi_h^{k+1} - \varphi_h^{k}), \qquad (5.87)$$

$$\Psi_-(\varphi_h^{k+1}) - \Psi_-(\varphi_h^{k}) \leq \Psi_-'(\varphi_h^{k})(\varphi_h^{k+1} - \varphi_h^{k}). \qquad (5.88)$$

The inequality is now obtained by testing (5.81) with μ_h^{k+1}, (5.82) with $(\varphi_h^{k+1} - \varphi_h^k)/\delta$, and adding the resulting equations. This leads to

$$\begin{aligned}
&\frac{1}{\delta}(\varphi_h^{k+1} - \varphi_h^{k}, \mu_h^{k+1}) + (m(\varphi_h^k)\nabla\mu_h^{k+1}, \nabla\mu_h^{k+1}) \\
&\qquad + \epsilon\frac{1}{\delta}(\nabla\varphi_h^{k+1}, \nabla(\varphi_h^{k+1} - \varphi_h^{k})) - \frac{1}{\delta}(\mu_h^{k+1}, \varphi_h^{k+1} - \varphi_h^{k}) \\
&\qquad + \frac{1}{\epsilon}\frac{1}{\delta}(\Psi_+'(\varphi_h^{k+1}), \varphi_h^{k+1} - \varphi_h^{k}) + \frac{1}{\epsilon}\frac{1}{\delta}\Psi_-'(\varphi_h^{k}), \varphi_h^{k+1} - \varphi_h^{k}) = 0.
\end{aligned}$$

Equality (5.86) and Inequalities (5.87) and (5.88) now imply

$$\int_\Omega m(\varphi_h^k)|\nabla \mu_h^{k+1}|^2\,dx + \frac{\epsilon}{2\delta}\int_\Omega |\nabla \varphi_h^{k+1}|^2\,dx - \frac{\epsilon}{2\delta}\int_\Omega |\nabla \varphi_h^k|^2\,dx$$

$$+\frac{\epsilon}{2\delta}\int_\Omega |\nabla \varphi_h^{k+1} - \nabla \varphi_h^k|^2\,dx + \frac{1}{\epsilon}\frac{1}{\delta}\int_\Omega \left(\Psi(\varphi_h^{k+1}) - \Psi(\varphi_h^k)\right)\,dx \le 0,$$

which yields the claim. □

Remark 5.10.6 For a fully practical finite element scheme one would need to use quadrature formulas for some of the integrals. Also for some of such schemes stability can be shown, see, e.g., [13].

Numerical Methods for Complex Interface Evolutions

6

Abstract

In this chapter we focus on the numerical aspects of complex interface problems. First, the two major approaches, interface capturing and interface tracking, are discussed and compared. Then, the application of mesh moving as the most straightforward representative of interface tracking as well as the level-set method as a prominent example of interface capturing to the numerical approximation of two phase flows are detailed.

6.1 Introduction and General Remarks About the Different Methods

The choice of how to computationally represent the interface within the numerical methodology for a free boundary problem may be viewed as the most important one, since it has significant implications for all subsequent steps in the development of the overall algorithm.

In general, one is faced with the following general difficulties:

- Discontinuities at the interface with possibly huge jumps in the pressure, in the concentration and in the gradients of the velocity.
- Geometry: efficient and reliable computation of curvature quantities, guarantee of volume conservation.
- In Marangoni flow the equations depend on *gradients* of temperature, concentration, etc.
- Strong nonlinear coupling. How can a stable time discretization be achieved?
- Efficient solution techniques. How to solve the arising systems of equations?

E. Bänsch et al., *Interfaces: Modeling, Analysis, Numerics*,
Oberwolfach Seminars 51, https://doi.org/10.1007/978-3-031-35550-9_6

Most numerical methods for moving boundary problems can be categorized into two fundamentally distinct classes: *Interface capturing* and *interface tracking* methods. In *interface capturing methods*, the interface is described implicitly by an additional scalar function. Prominent representatives of this method are volume of fluid (VOF) methods [90, 136], phase field approaches [50,56,96] and level set methods [87, 124, 135, 141, 144]. The latter is the by far most common approach in two phase flow.

Extensions of these methods in order to incorporate evaporation/condensation are treated for instance in [81, 143, 145, 153] for the level set method, in [137, 149] for VOF and for the phase field method in [97, 127].

In contrast to interface capturing methods, in *interface tracking* the phase boundary is described explicitly in terms of the computational mesh. Consequently, the computational mesh deforms according to the interface motion, see for example [10,49,91]. While usually limited to situations in which only moderate deformations and no topological changes of the interface occur, the explicit representation of the interface allows for a very accurate treatment of surface tension. Interface tracking is thus the method of choice in those situations, where only moderate deformations occur and the topology of the interface stays fixed.

Interface capturing

- PROS:
  - Very general tools, no problem with changes of topology.
  - Level set and phase field methods seem to become *the* standard tools (competition with VOF).
- CONS:
  - How to compute curvature (see for instance Groß and Reusken [87] and Reusken and Esser [132] for level set methods)?
  - Discontinuities: use XFEM in level set methods to e.g. avoid spurious velocities (serious issue), often overlooked. In phase field methods discontinuities are smeared out.
  - Volume conservation.

Interface tracking

- CONS:
  - Mesh moving methods are limited to moderate interface deformations ($\approx$50%).
  - Topological changes nearly impossible to handle (can be done, see, e.g., [25,26,103], but very painful).
  - Moving meshes can lead to small cells. However, unfitted methods can deal with this but they are less accurate.

- PROS:
  - Simple and most accurate method.
  - No problems with geometric quantities like curvature etc.
  - No problems with discontinuities with the moving mesh method.

Upshot: choice of method depends on the specific problem to be solved!
For more details on numerical methods for interfaces we refer to [11,23,46,50,54].

6.2 Interface Capturing

6.2.1 Level Set Methods

One of the most popular interface capturing methods is the level set method by Osher and Sethian [123, 124]. Here, Γ is given as a level set (without loosing generality the 0 level set) of the *level set function* ϕ:

$$\phi : [0, T] \times \Omega \to \mathbb{R}$$

and $\Gamma(t) = \{x \in \Omega \mid \phi(t, x) = 0\}$.

- PROS:
  - Very general and versatile method. (Nearly) no problems with large deformations and topology changes.
  - Topology changes come for free, no extra effort necessary.
- CONS:
  - More effort for developing algorithms and implementation.
  - Error analysis is not easy, little analytical results are known for bulk-interface coupling.
  - One has to solve in one dimension higher.
  - Computing geometric quantities is painful.
  - Often problems with mass conservation appear. However, see the work of Kees et al. [101] and Quezada de Luna et al. [130] who were able to deal with this problem.
  - "Nice on the paper", but to make it work neatly, a lot of nasty details have to be taken into account.

Nevertheless, level set methods are by far the most popular methods for complicated two phase problems etc.

6.2.2 Phase Field Methods

In a sense, phase field methods are different than the other methods mentioned in that they regularize the problem and make all the variables smooth. This, of course has several advantages, both from an analytical point of view as well as for numerics. Moreover, interface conditions in phase field methods are *physics based* and allow for a relatively easy incorporation of many physical phenomena.

The drawback of phase field methods is the demand to sufficiently resolve the smeared interfacial layer of order ε, cf., Sects. 3.8 and 5.9 for the role of the parameter ε. Depending on the problem or application at hand this may result in prohibitively fine meshes. For certain applications this could require orders of magnitude more degrees of freedom (DOF) than for instance with a sharp interface/mesh moving method. Nevertheless, if for the problem to be solved this is not the key issue, then phase field methods can be very strong numerical tools.

Since there is no explicit free boundary and all variables are smooth, the numerical approximation of a phase field problem consists in approximating a system of (second or fourth order in space) PDEs. Time discretization can take advantage of the convex–concave structure of the energy functional: convex parts are solved implicitly while the nasty concave parts are treated explicitly, see Sect. 5.10.1 for details. One advantage is that one can use standard finite elements to discretize in space. Disadvantages are that error estimates typically depend exponentially on ε^{-1} and that one has to compute the phase field instead of a parametrization which means that one has to solve in one dimension higher. In addition, it is not easy to resolve small distances between interfaces and boundaries.

6.3 Interface Tracking

Denote by $\Gamma(t)$ the evolving interface that is explicitly represented in interface tracking. We distinguish between two main classes: *mesh moving* and *front tracking*.

6.3.1 Mesh Moving Approaches (Fitted Approaches)

Let $\Omega_1(t)$, $\Omega_2(t)$ denote for instance two phases that are separated by the sharp and smooth interface $\Gamma(t)$. Computationally, we have $\Omega_{1,h}(t_k)$, $\Omega_{2,h}(t_k)$ at discrete time instants t_k. Now mesh moving means that the two phases are separated by $\Gamma_h(t_k)$, which is the boundary of the triangulations both of $\Omega_{1,h}(t_k)$ and $\Omega_{2,h}(t_k)$. The joint triangulation is a conforming triangulation of the overall domain Ω. The discrete interface $\Gamma_h(t_k)$ is thus a conforming $(n-1)$-dimensional triangulation approximating the exact interface. This means that the discrete interface is *fitted* to the discrete bulk domains. In the mesh moving or fitted mesh approach, the discrete interface is hence made up of faces of elements from

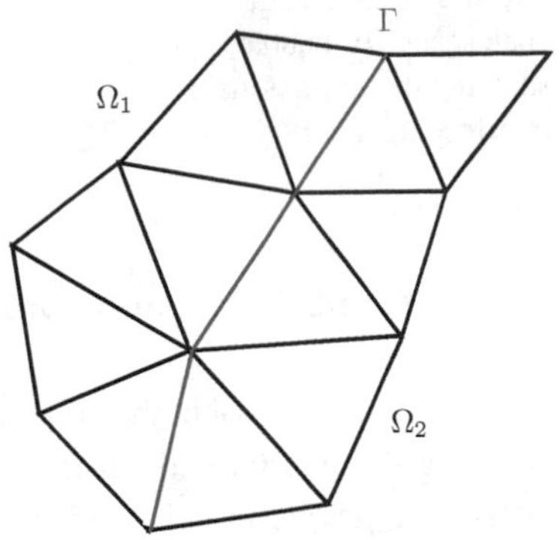

Fig. 6.1 Mesh moving method

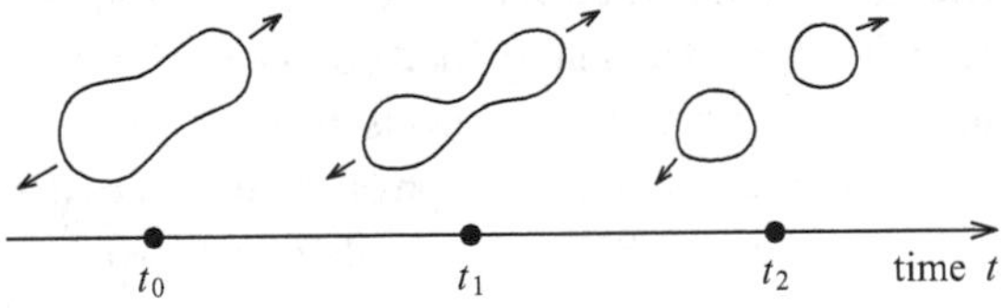

Fig. 6.2 Example of breakup of bubbles showing the limitations of interface tracking

the bulk mesh, and thus the bulk mesh needs to deform appropriately in time in order to match the evolving interface. Employing so-called Eulerian schemes on this moving bulk mesh implies that the obtained solutions need to be frequently interpolated on the new mesh. In purely Lagrangian schemes, on the other hand, the bulk mesh needs to deform according to the fluid velocity, and this often leads to large distortions of the mesh. A possible way to overcome these difficulties is the arbitrary Lagrangian–Eulerian (ALE) approach, e.g., [91], where the equations are formulated in a moving frame of reference, and the corresponding reference velocity is independent from the fluid velocity, and in this sense it is "arbitrary" (Fig. 6.1).

- PROS:
  - Very precise representation of all geometric quantities, in particular if higher order elements (curved simplices) are used for triangulating the two phases.
  - Handling of discontinuities rather straightforward.
  - Very simple to implement—as long as no remeshing is necessary.
  - Very small numerical overhead.
  - Small distances between interfaces and boundaries can be resolved.
- CONS:
  - In some cases a reassembly is needed in every time step.
  - Works only for moderate mesh deformations.
  - Topology changes are extremely painful to handle (see Figs. 6.2 and 6.3).

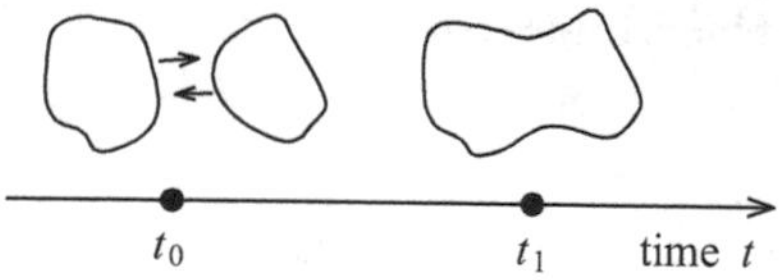

Fig. 6.3 Example of coalescence of bubbles showing the limitations of interface tracking

6.3.2 Front Tracking Approaches (Unfitted Approaches)

In the unfitted approach, the bulk mesh and the interface mesh are totally independent, thus allowing the interface to cut through the elements of the bulk mesh. The computational interface $\Gamma_h(t_k)$ is an independent $(n-1)$-dimensional triangulation that moves through Ω according to its evolution law. One of the prominent examples for unfitted approximations is the immersed boundary method, see [107, 126]. In the finite element framework, usually an enrichment of the elements that are cut by the interface is necessary in order to accurately capture jumps of physical quantities across the interface, see [9, 19] for XFEM-approaches and [38] for the cutFEM-approach.

- PROS:
  - Allows for larger deformations than mesh moving (think of a rising bubble in a surrounding liquid).
  - Very precise representation of all geometric quantities.
  - Very simple to implement.
  - Very small numerical overhead.
  - Small distances between interfaces and boundaries can be resolved.
- CONS:
  - Loss of accuracy, especially in handling discontinuities.
  - Something has to be done to avoid *spurious velocities*, see below.
  - "Exchange of information" between bulk mesh and interface through surface integrals needs to be computed.

Remark 6.3.1 (Spurious Velocities) In incompressible flow problems all methods except for mesh moving have difficulties with discontinuous pressures. To understand this problem, we consider a *static* situation, $u \equiv 0$, of a two phase flow problem as depicted in Fig. 6.4. In this case, the Navier–Stokes equations and the interface condition reduce to

$$\nabla p = 0 \quad \text{in } \Omega_1, \Omega_2, \qquad [p] := p_2 - p_1 = \gamma\kappa = \gamma\frac{n-1}{R} \quad \text{on } \Gamma.$$

(continued)

This means the pressure is piecewise constant but experiences a (possibly huge) jump.

If the numerical method does not allow for such a pressure jump (for instance, because the interface intersects an element), the numerical pressure p_h will fulfill $\nabla p_h \neq 0$ in Ω_1, Ω_2 in general. Then in turn, $u_h \neq 0$ (see Fig. 6.5). This has to be cured by some means, for instance XFEM, see below.

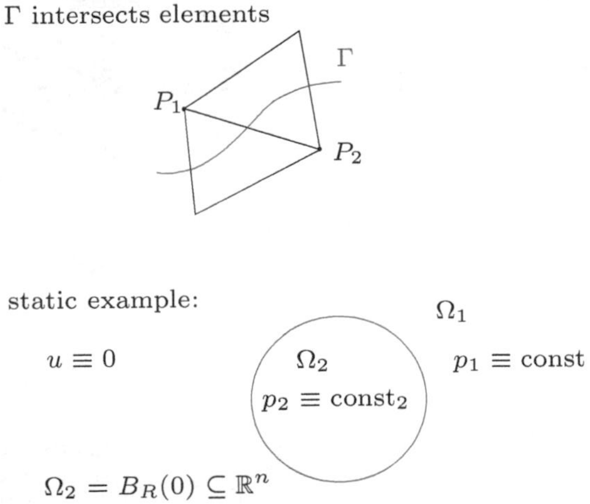

Fig. 6.4 Static example

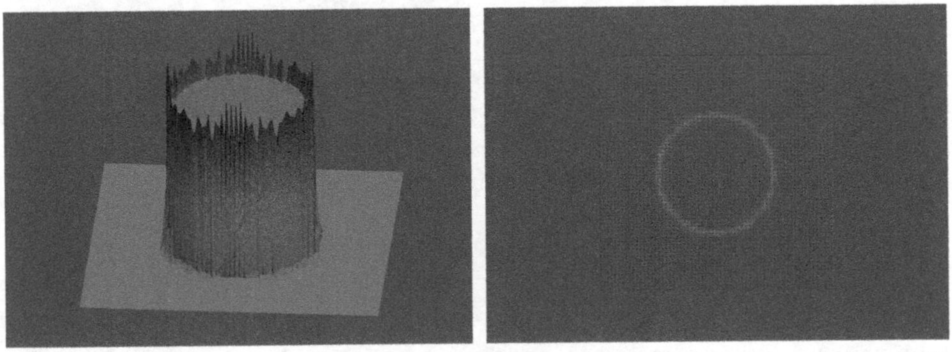

Fig. 6.5 Static example as above computed with the Taylor–Hood element (piecewise quadratics for the velocity, piecewise linears for the pressure) without XFEM. Left: pressure; right: magnitude of spurious velocity

6.4 Two Phase Flow

We restate the equations describing the motion of two viscous, incompressible fluids with an interface taking surface tension into account. The equations derived in Sect. 3.7, in the case where the interface does not intersect the outer boundary, are as follows:

$$\rho_i \cdot (\partial_t u + u \cdot \nabla u) - \mu_i \Delta u + \nabla p = f \qquad \text{in } \Omega_i(t), \qquad i = 1, 2, \tag{6.1}$$

$$\nabla \cdot u = 0 \qquad \text{in } \Omega_i(t) \tag{6.2}$$

where on $\Gamma(t)$ we prescribe

$$[u]_{l_1}^{l_2} = 0, \tag{6.3}$$

$$[2\mu D(u)\nu - p\nu]_{l_1}^{l_2} = -\gamma\kappa\nu - \nabla_\Gamma \gamma, \tag{6.4}$$

$$u \cdot \nu = V. \tag{6.5}$$

This system has to be solved together with initial conditions for u and Γ:

$$u(0, \cdot) = u_0, \qquad \Gamma(0) = \Gamma_0. \tag{6.6}$$

As boundary conditions for u on the outer boundary $\partial\Omega$ we take homogeneous Dirichlet conditions.

We will now describe how mesh moving and level set methods can be used to solve the above problem numerically.

6.4.1 Mesh Moving

Arbitrary Lagrangian–Eulerian Coordinates (ALE)

As already outlined, in mesh moving the discrete interface is explicitly tracked by a (discrete) evolution law, see Fig. 6.6 for an example. The shift of the DOFs at the interface is a first order in time discretization of the kinematic boundary condition Eq. (6.5).

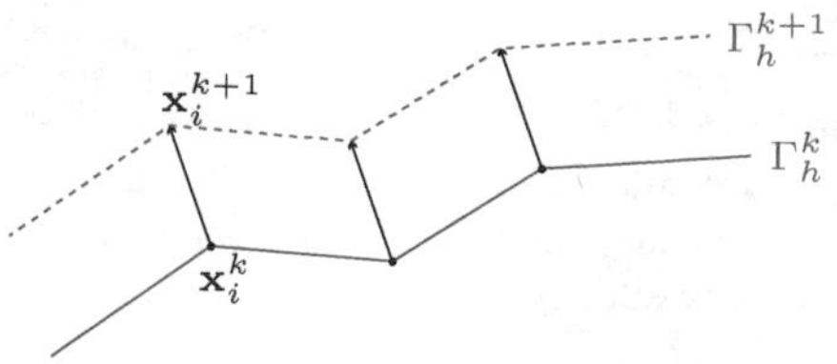

Fig. 6.6 Mesh moving method: defining the new position of the interface. The mesh points at the interface are moved with the fluid velocity

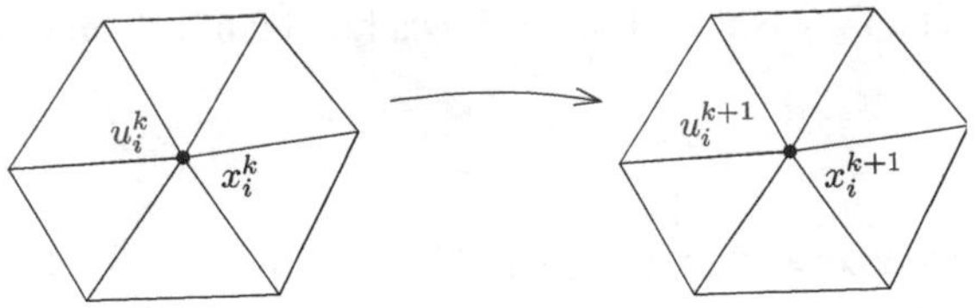

Fig. 6.7 Information traveling with the mesh

To avoid strong mesh distortion, this movement has to be extended and smoothed out to the bulk meshes for either phase $\Omega_{1,h}$, $\Omega_{2,h}$ by an extension operator in such a way that the mesh topology is preserved. As extension operator one can use the discrete Laplace operator or more sophisticated operators like operators from linear or nonlinear elasticity.

Since the location of the DOFs (for instance the vertices) carry the information of finite element functions, this information is also moved in turn (Fig. 6.7). To compensate for this movement of information the corresponding *mesh velocity* has to be added to the transport equations. This methodology is called ALE, see for instance [10, 49, 91]. We will explain this concept next.

Let the finite element functions, for instance the velocity, be given by

$$u_h^k(x) = \sum_i u_i^k \varphi_i^k(x) \approx u(t_k, x),$$

where the $\{\varphi_i^k\}$ are the finite element basis functions corresponding to the mesh at time t_k.

This shows that in general (assuming for simplicity that we use a backward Euler scheme with a time step δ)

$$\sum_i \frac{(u_i^{k+1} - u_i^k)\varphi_i^k}{\delta} \not\approx \partial_t u(t_{k+1}, \cdot).$$

At first glance, ALE may appear to be a bit complicated. Later it will (hopefully) become clear that this approach is rather simple and very easy to implement. In fact, only one additional term has to be added to the advection terms.

Eulerian Coordinates

These are given by the frame of a fixed observer. Usually one wants to have the PDEs in this frame and for the transport part of the momentum equation for instance one gets

$$\rho(\partial_t u + u \cdot \nabla u) + \ldots = \ldots.$$

Lagrangian Coordinates

Usually one derives the Navier–Stokes equations from considering a fixed, time independent *reference domain* $\hat{\Omega}$, for instance the volume that is occupied by the fluid initially.

The flow is then described by a family of diffeomorphisms $\Phi(t, \cdot)$,

$$\Phi(t, \cdot) : \hat{\Omega} \to \Omega(t).$$

Now $x = \Phi(t, \hat{x})$ denotes the position of the "fluid particle" being at location $\hat{x}$ in the reference domain $\hat{\Omega}$ which could be, as discussed before, just the initial domain. We also refer to the notion of characteristic curves in Sect. 6.4.2. The point $\hat{x}$ is called the *Lagrangian coordinate* of a point x at time t. The velocity u in turn is defined by

$$u(t, x) := \frac{d}{dt} x(t) = \partial_t \Phi(t, \hat{x})$$

with $\hat{x} := \Phi^{-1}(t, x) \in \hat{\Omega}$ and the path line $x(t) = \Phi(t, \hat{x})$ which is the path the material point $\hat{x}$ will follow during the evolution. Here, and in what follows, the inverse is taken with respect to the second variable.

Let α be an arbitrary quantity transported by the flow. Then

$$\frac{d}{dt}\alpha(t, x(t)) = \partial_t \alpha(t, x(t)) + \frac{d}{dt} x(t) \cdot \nabla \alpha(t, x(t)) = (\partial_t \alpha + u \cdot \nabla \alpha)(t, x(t)).$$

The expression

$$D_t \alpha = \partial_t \alpha + u \cdot \nabla \alpha \tag{6.7}$$

is the *material time derivative* from Sect. 2.8.

ALE Coordinates

Let $\tilde{\Omega}$ be a further reference domain and

$$\Psi(t, \cdot) : \tilde{\Omega} \to \Omega(t)$$

a further diffeomorphism. The map $\Psi(t, \tilde{x})$ does not necessarily have to follow the flow, i.e., the mapping $t \mapsto \Psi(t, \tilde{x})$ does not necessarily describe the path of the fluid particle $\tilde{x}$. We now define the velocity

$$w(t, x) := \partial_t \Psi(t, \tilde{x}) \qquad \text{for } \tilde{x} = \Psi^{-1}(t, x). \tag{6.8}$$

Later w will be the *mesh velocity*.

Remark 6.4.1 Let $\hat{\Omega} = \tilde{\Omega} = \Omega(0)$. Then we can consider the following cases.

1. $w \equiv 0$ $\Leftrightarrow$ $\tilde{x} = x$, i.e. $\tilde{x}$ are the Eulerian coordinates.
2. $w \equiv u$ $\Leftrightarrow$ $\tilde{x} = \hat{x}$, i.e. $\tilde{x}$ are the Lagrangian coordinates.
3. $w \not\equiv 0, w \not\equiv u$: corresponds to a general arbitrary Lagrangian-Eulerian (ALE) setting.

Figure 6.8 shows an illustration of the tree types of coordinates. The connection between the material time derivative and the time derivative in ALE coordinates is stated in the following proposition.

Fig. 6.8 Eulerian, Lagrangian and ALE coordinates

Proposition 6.4.2 *Let* $\alpha : [0, T] \times \Omega \to \mathbb{R}$ *be smooth and* $\tilde{\alpha}(t, \tilde{x}) := \alpha(t, \Psi(t, \tilde{x}))$. *Then it holds*

$$D_t\alpha(t, x) = \partial_t\tilde{\alpha}(t, \tilde{x}) + \Big(u(t, x) - w(t, x)\Big) \cdot \nabla_x\alpha(t, x)$$

with $x = \Psi(t, \tilde{x})$.

Proof Using the chain rule we compute

$$\begin{aligned}
\partial_t\tilde{\alpha}(t, \tilde{x}) &= \frac{d}{dt}\alpha(t, \Psi(t, \tilde{x})) \\
&= \partial_t\alpha(t, \Psi(t, \tilde{x})) + \nabla_x\alpha(t, \Psi(t, \tilde{x})) \cdot \partial_t\Psi(t, \tilde{x}) \\
&= \partial_t\alpha(t, x) + \nabla_x\alpha(t, x) \cdot w(t, x) \\
&= D_t\alpha(t, x) - \nabla_x\alpha(t, x) \cdot u(t, x) + \nabla_x\alpha(t, x) \cdot w(t, x)
\end{aligned}$$

where we used Eq. (6.8) and the definition of the material time derivative, see (6.7). □

Moving Mesh Method for the Two Phase Problem

Now we have all the pieces together to define the generic form of the mesh moving method. First of all we recall the weak formulation of two-phase flow presented in Definition 3.7.1. It is this weak formulation which we are going to discretize in what follows.

In fact in the weak formulation we searched for a pair (u, p) with $(u(t), p(t)) \in X \times Y$ for all $t \in [0, T]$ and an evolving hypersurface Γ such that $u(0, \cdot) = u_0$, $\Gamma(0) = \Gamma_0$ and fulfilling for all $t \in [0, T]$

$$(\rho(\partial_t u + u \cdot \nabla u), \varphi) + (2\mu D(u), D(\varphi)) - (p, \nabla \cdot \varphi) + \int_{\Gamma(t)} \gamma \nabla_{\Gamma(t)} \cdot \varphi = 0 \qquad \forall \varphi \in X, \tag{6.9}$$

$$(\nabla \cdot u, q) = 0 \qquad \forall q \in Y. \tag{6.10}$$

Using Proposition 6.4.2 with $\alpha = u$ we may rewrite the first term in (6.9) as

$$\begin{aligned}
(\rho(\partial_t u + u \cdot \nabla u), \varphi) &= \int_\Omega \rho(t, x) D_t u(t, x) \cdot \varphi(x) dx \\
&= \int_\Omega \rho(t, x) \Big[\partial_t \tilde{u}(t, \Psi^{-1}(t, x)) \\
&\qquad + (u(t, x) - w(t, x)) \cdot \nabla u(t, x)\Big] \cdot \varphi(x)) dx,
\end{aligned}$$

where $\tilde{u}(t, \tilde{x}) = u(t, \Psi(t, \tilde{x}))$. The time derivative $\partial_t \tilde{u}(t, \Psi^{-1}(t, x))$ for simplicity is discretized with the help of the backward Euler method. In fact we take

$$\begin{aligned}\partial_t \tilde{u}(t, \Psi^{-1}(t, x)) &\approx \frac{1}{\delta}\Big(\tilde{u}(t, \Psi^{-1}(t, x)) - \tilde{u}(t-\delta, \Psi^{-1}(t, x))\Big)\\ &= \frac{1}{\delta}\Big(u(t, x) - \tilde{u}(t-\delta, \Psi^{-1}(t, x))\Big).\end{aligned}$$

Hence the difference quotient used to approximate the time derivative involves points with the same ALE spatial coordinates. In the algorithm the ALE coordinates evolve via the grid velocity. In practice one moves the grid and computes the difference quotient involving values at the old and the new spatial grid points.

Here we used the backward Euler scheme for simplicity. Using the above weak formulation we are going to formulate an algorithm for mesh-moving for two-phase flow. Recall that the whole flow domain Ω is time independent and that $\Omega_h = \Omega$. Furthermore, we assume the densities ρ_i and viscosities μ_i to be piecewise constant in either phase $i = 1, 2$.

Algorithm 6.4.3 (Mesh Moving for Two Phase Flow) *Let Γ_h^0, $\Omega_{1,h}^0$, $\Omega_{2,h}^0$, $u_h^0 = \tilde{u}_h^0$ and $w_h^0 = 0$ be given. Set $\rho_h^0 = \rho_i$ and $\mu_h^0 = \mu_i$ in $\Omega_{i,h}^0$, $i = 1, 2$.*

In each discrete time step $t_k = k\delta$, $k = 0, 1, 2, \ldots$, do the following steps.

1. *Find u_h^{k+1}, p_h^{k+1}, X^{k+1} fulfilling for all $\varphi_h \in X_h^k$ and $q_h \in Y_h^k$*

$$(\rho_h^k \frac{u_h^{k+1} - \tilde{u}_h^k}{\delta}, \varphi_h) + (\rho_h^k (u_h^{k+1} - w_h^k)\cdot \nabla u_h^{k+1}), \varphi_h) - (p_h^{k+1}, \nabla \cdot \varphi_h)$$

$$+ (2\mu_h^k D(u_h^{k+1}), D(\varphi_h)) + \int_{\Gamma_h^*} \gamma \nabla_{\Gamma_h^*} \cdot \varphi_h d\mathcal{H}^{n-1} = 0,$$

$$(\nabla \cdot u_h^{k+1}, q_h) = 0,$$

$$X^{k+1} = id_{\Gamma_h^k} + \delta\, u^{k+1}_{h\,|\Gamma_h^k}.$$

Possible choices of Γ_h^* are discussed below.

2. *Extend X^{k+1} to the interior of the fluid phases $\Omega_{1,h}^k$ and $\Omega_{2,h}^k$ to a mapping Υ^{k+1} such that*

$$\Omega_{1,h}^{k+1} = \Upsilon^{k+1}(\Omega_{1,h}^k) \quad \text{and} \quad \Omega_{2,h}^{k+1} = \Upsilon^{k+1}(\Omega_{2,h}^k).$$

(continued)

The degrees of freedom of the computational grid are moved by Υ^{k+1}. A possible way to compute Υ^{k+1} is discussed below.
Set $\rho_h^{k+1} = \rho_i$ and $\mu_h^{k+1} = \mu_i$ in $\Omega_{i,h}^{k+1}$, $i = 1, 2$.

3. *Compute the grid velocity w_h^{k+1} on the new mesh by*

$$w_h^{k+1} \circ \Upsilon^{k+1} := \frac{\Upsilon^{k+1} - id}{\delta}.$$

Remark 6.4.4

- In the above algorithm $\tilde{u}_h^k$ is the old velocity u_h^k on $\Omega_{i,h}^{k-1}$ but lifted to $\Omega_{i,h}^k$, $i = 1, 2$, i.e.,

$$\tilde{u}_h^k = u_h^k \circ (\Upsilon^k)^{-1},$$

where $(\Upsilon^k)^{-1}$ is the inverse of Υ^k. Numerically there is nothing to do, since we work with the same DOFs in the old and new mesh.
- The discrete interface Γ_h^* denotes either Γ_h^k or Γ_h^{k+1}, see the discussion below.
- The finite element space Y_h^k for the pressure is defined in such a way that pressure nodes at the interface are virtually *doubled*, i.e. there exist two copies of a pressure node at the interface, one belonging to $\Omega_{1,h}$ the other one to $\Omega_{2,h}$ (Fig. 6.9). This allows the pressure to exhibit a jump across the interface and therefore no spurious velocities occur. The finite element space X_h^k for the velocity needs to be chosen in such a way that (X_h^k, Y_h^k) fulfill the Ladyzhenskaya-Babuska-Brezzi (LBB) stability condition. For example the Taylor-Hood elements are a possible choice.

Remark 6.4.5

- There are situations, where a *full update*, i.e. shifting the interface nodes with the full velocity,

$$X^{k+1} = id_{\Gamma_h^k} + \delta\, u_{h\,|\Gamma_h^k}^{k+1},$$

(continued)

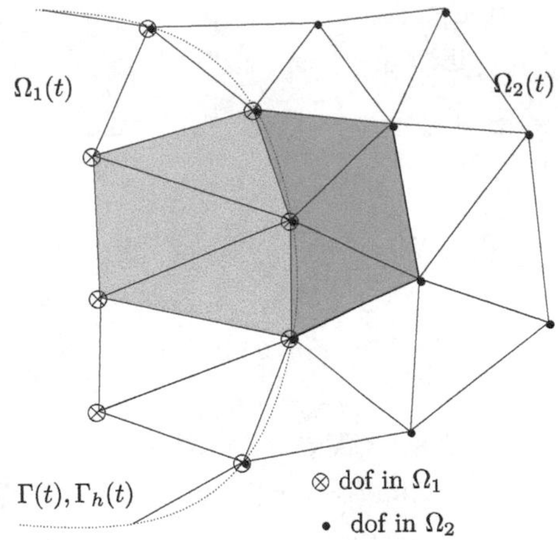

Fig. 6.9 The support of a pressure basis function ψ_i lies only in either phase

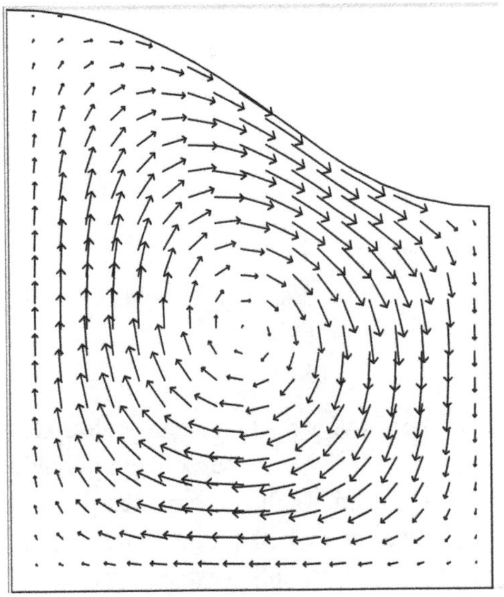

Fig. 6.10 Free surface flow: upper surface is the free surface with a strong tangential flow field due to Marangoni convection. A *full update* would quickly cluster the boundary nodes at the upper right corner leading to complete distortion of the mesh

leads to a quick distortion of the mesh, see Fig. 6.10. In this case one would rather use an update only in normal direction. Since the discrete normals to the discrete interface are in general discontinuous at vertices (or edges in 3d), one may use smoothed or averaged normals $\tilde{\nu}$ that are obtained by a weighted average of the elementwise discrete normals, where the weight is according to the length (or area in 3d) of the interface edge (or face in 3d). With the help of these averaged normals the kinematic boundary condition (6.5) is discretized by a first order Euler approximation, i.e. the Lagrange nodes on the interface x_i^k are shifted to

(continued)

their new positions by

$$x_i^{k+1} = x_i^k + \delta \left(u_h^{k+1}(x_i) \cdot \tilde{\nu}(x_i) \right) \tilde{\nu}(x_i),$$

which can be also written in terms of the Lagrange interpolation operator I_h as

$$X^{k+1} = id_{\Gamma_h^k} + \delta I_h \left((u_{h\,|\Gamma_h^k}^{k+1} \cdot \tilde{\nu})\, \tilde{\nu} \right)$$

with $X^{k+1}(x_i^k) = x_i^{k+1}$.

- A more general approach to updating the free boundary is the use of a given update direction $z_i^k \in \mathbb{R}^n$ for each Lagrange node x_i^k of the free surface Γ_h^k. Then the update for x_i^k is defined by

$$x_i^{k+1} = x_i^k + \delta\, \eta_i z_i^k,$$

where η_i has to fulfill $(\eta_i z_i^k) \cdot \tilde{\nu}_i = u_h^{k+1} \cdot \tilde{\nu}_i$ and is thus given by

$$\eta_i = u_h^{k+1}(x_i^k) \cdot \tilde{\nu}_i / (z_i^k \cdot \tilde{\nu}_i).$$

Of course, this requires that $z_i^k \cdot \tilde{\nu}_i \neq 0$.

In the above algorithm the mesh is updated from time step to time step by an extension Υ^{k+1} of X^{k+1}. We now describe a possible way of doing so. For each $k = 0, 1, 2, 3 \ldots$ we now assume we have computed the polyhedral surface Γ_h^k. Let x_i^k, $i = 1, \ldots, k_\Omega$, be the vertices of the bulk mesh at time step t_k. We assume that the vertices of Γ_h^k are part of the vertices of the bulk mesh and denote by $\Omega_{1,h}^k$ and $\Omega_{2,h}^k$ the discrete approximations of the two phases which are separated by Γ_h^k.

The vertices of the mesh are updated according to

$$x_i^{k+1} = \Upsilon^{k+1}(x_i^k) \qquad \text{for} \qquad i = 1, \ldots, k_\Omega, \tag{6.11}$$

where Υ^{k+1} is the displacement of the bulk mesh which we are going to specify now. With the help of

$$W_h^k = \{\chi \in C^0(\overline{\Omega}, \mathbb{R}^n) \mid \chi \text{ piecewise linear, } \chi \cdot n = 0 \text{ on } \partial\Omega,\, \chi = X^{k+1} \text{ on } \Gamma_h^k\},$$

$$W_{h,0}^k = \{\chi \in C^0(\overline{\Omega}, \mathbb{R}^n) \mid \chi \text{ piecewise linear, } \chi \cdot n = 0 \text{ on } \partial\Omega,\, \chi = 0 \text{ on } \Gamma_h^k\},$$

we compute $\Upsilon^{k+1} \in W_h^k$ such that

$$(\nabla \Upsilon^{k+1}, \nabla \chi_h) = 0 \tag{6.12}$$

holds for all $\chi_h \in W_{h,0}^k$, which is an extension by the (discrete) Laplacian. Sometimes working with the (discrete) operator of linear elasticity gives better results. In this case, Υ^{k+1} is given by

$$2(D(\Upsilon^{k+1}), D(\chi_h)) + (\lambda \nabla \cdot \Upsilon^{k+1}, \nabla \cdot \chi_h) = 0 \tag{6.13}$$

for all $\chi_h \in W_{h,0}^k$ with $\lambda > 0$.

The simplest choice is a constant λ. However, one can also choose λ depending on x which can help to improve the mesh quality, see, e.g., [80]. An even more sophisticated approach is described in [133], where an operator motivated by nonlinear elasticity is used.

The discrete mesh velocity is then defined by

$$w_h^{k+1} = \sum_{k=1}^{k\Omega} \left(\frac{x_i^{k+1} - x_i^k}{\delta} \right) \varphi_i^{k+1},$$

where φ_i^{k+1} are the nodal basis functions corresponding to the new mesh points x_i^{k+1}. We now obtain

$$(\Upsilon^{k+1})^{-1} = \mathrm{id} - \delta\, w_h^{k+1}$$

and observe

$$(\Upsilon^{k+1})^{-1}(\Omega_{i,h}^{k+1}) = \Omega_{i,h}^k \quad \text{for} \quad i = 1, 2.$$

We set $\widetilde{u}_h^{k+1} = u_h^{k+1} \circ (\Upsilon^{k+1})^{-1}$ which then can be used to compute the discretization of the ALE time derivative $\partial_t \widetilde{u}$ in the next time step (see the remark above).

Let $\gamma = const$. The curvature term in the above algorithm

$$e_*(\varphi_h) = \int_{\Gamma_h^*} \gamma \nabla_{\Gamma_h^*} \cdot \varphi_h d\mathcal{H}^{n-1}$$

can be reformulated with the help of the relation

$$\nabla_{\Gamma_h^*} \cdot \varphi_h = \nabla_{\Gamma_h^*} id : \nabla_{\Gamma_h^*} \varphi_h.$$

Here, one uses that $\nabla_{\Gamma_h^*} id = P_h^*$, where P_h^* is the projection on the tangent space of Γ_h^*, and hence $\nabla_{\Gamma_h^*} \cdot \varphi_h = Tr(\nabla_{\Gamma_{h*}} \varphi_h) = Tr((\nabla_{\Gamma_h^*} \varphi_h) P_h^*) = \nabla_{\Gamma_h^*} id : \nabla_{\Gamma_h^*} \varphi_h$. Therefore, we obtain

$$e_*(\varphi_h) = \int_{\Gamma_h^*} \gamma \nabla_{\Gamma_h^*} id : \nabla_{\Gamma_h^*} \varphi_h d\mathscr{H}^{n-1}.$$

It remains to determine $*$. If we take $* = k$, then all geometry terms in the algorithm are treated explicitly. Thus the computation of the flow field and the geometry is completely decoupled. However, in this case one faces a severe CFL-like condition for stability. In fact the time step δ needs to fulfill

$$\delta \leq \frac{C}{\sqrt{\gamma}} h^{3/2},$$

see for instance [10,30]. The choice $* = k+1$ results in an unconditionally stable scheme, but now the geometry and the flow field are coupled (we need Γ_h^{k+1} to compute u_h^{k+1}).

There is an alternative, decoupling geometry and flow field, but being unconditionally stable at the same time. In fact one chooses

$$e_{semi}(\varphi_h) = \int_{\Gamma_h^k} \gamma \nabla_{\Gamma_h} X^{k+1} : \nabla_{\Gamma_h} \varphi_h d\mathscr{H}^{n-1},$$

where $X^{k+1}$is given by

$$X^{k+1}(z) := z + \delta u_h^{k+1}(z) \qquad \text{for } z \in \Gamma_h^k.$$

This leads to

$$e_{semi}(\varphi_h) = \int_{\Gamma_h^k} \gamma \nabla_{\Gamma_h^k} (id + \delta u_h^{k+1}) : \nabla_{\Gamma_h^k} \varphi_h d\mathscr{H}^{n-1}.$$

This scheme is semi-implicit in the curvature. There is an implicit contribution to the computation of u_h^{k+1}. However, the geometry is still explicit. Unconditional stability was proved (for slightly different settings) in [10, 80]. See also [150, 151] for the fully implicit approach $* = k + 1$ and further time discretizations.

6.4.2 Level Set Method for Two Phase Flow

Let $\Omega_1(t)$, $\Omega_2(t)$ denote the two phases separated by the sharp interface $\Gamma(t)$. As described above, $\Gamma(t)$ is given as the zero level of the level set function ϕ:

$$\Gamma(t) := \{x \in \Omega \mid \phi(t, x) = 0\}.$$

Additionally we assume

$$\phi(t,\cdot) < 0 \qquad \text{in } \Omega_1(t),$$
$$\phi(t,\cdot) > 0 \qquad \text{in } \Omega_2(t).$$

We want to find an evolution law for ϕ. Assume that u is smooth. We exploit the fact that Γ is a *material* boundary, i.e. each phase is transported by the flow field u. We construct ϕ in such a way that it is constant on *characteristic curves* $t \mapsto \xi(t,\bar{x})$, that is:

$$\frac{d}{dt}\xi(t,\bar{x}) = u(t,\xi(t,\bar{x})) \qquad \text{for } t > 0,$$
$$\xi(0,\bar{x}) = \bar{x} \in \Omega.$$

On the characteristic curves the level set functions should be constant:

$$\frac{d}{dt}\phi(t,\xi(t,\bar{x})) = 0.$$

Using the chain rule and the definition of the characteristic curve, this is equivalent to the following transport equation:

$$\partial_t\phi(t,x) + u(t,x)\cdot\nabla\phi(t,x) = 0 \tag{6.14}$$

for $x = \xi(t,\bar{x})$. If the characteristic curves cover all of Ω, then the above equation holds for all $x \in \Omega$.

To close the description of ϕ, initial as well as boundary conditions on the inflow part of the boundary are needed:

$$\phi(0,\cdot) = \phi_0 \qquad \text{in } \Omega,$$
$$\phi = \phi_D \qquad \text{on } \partial_{in}\Omega_T$$

with ϕ_0, ϕ_D given and $\partial_{in}\Omega_T$ being the inflow boundary:

$$\partial_{in}\Omega_T := \{(t,x) \in [0,T]\times\partial\Omega \mid u(t,x)\cdot\nu(x) < 0\}.$$

Equation (6.14) together with the above initial and boundary conditions can be discretized by a stabilized finite element method, for instance by the SUPG method, see [34,87,99] for a few out of an abundance of possible references. Many other choices might be used as well.

It turns out that computations tend to break down, when $\nabla\phi$ becomes too large or close to 0. Thus we ideally want $\phi(t,\cdot)$ to be the signed distance function to $\Gamma(t)$ implying that

$$|\nabla\phi(t,\cdot)| = 1$$

at least close to the interface $\Gamma(t)$.

Even if ϕ is initially a signed distance function, this property is lost during the time evolution. Thus to regain this property, one usually incorporates a step in the algorithm that tries to make ϕ a distance function again each time step or every few time steps. This step is called *re-distancing* and may take a significant amount of computing time. There are many different methods for re-distancing, see for instance [65, 130, 131, 134, 139, 140] for a by far not complete list.

Re-distancing usually leads to another problem on the discrete level: The level set function is changed and then in turn also the numerical interface changes slightly, which may yield an error in the volume of either phase. This effect can accumulate over time to an unacceptable amount. Thus methods for correcting the volume have to be used.

Next we will describe how to incorporate the surface force

$$-\gamma \int_\Gamma \kappa\varphi \cdot \nu d\mathcal{H}^{n-1}$$

for test functions φ in the velocity space.

Since curvature is a second order operator on Γ and the numerical level set is continuous only, one cannot directly use the formula from Proposition 2.3.3. The remedy is to use integration by parts, see Chap. 2.

As for mesh moving one can use (assuming for the moment that Γ is closed) Eq. (3.56)

$$\int_\Gamma \kappa\varphi \cdot \nu\, d\mathcal{H}^{n-1} = \int_\Gamma \nabla_\Gamma \cdot \varphi d\mathcal{H}^{n-1}$$

or as above

$$e_*(\varphi) = \int_\Gamma \gamma \nabla_\Gamma id : \nabla_\Gamma \varphi d\mathcal{H}^{n-1}.$$

Both formulas require only the evaluation of first order derivatives on the test function φ and in the second case on the identity mapping on Γ and are thus well defined also in the discrete case.

Let us have a look how this can be done for a computational level set function ϕ_h. We first treat the case when ϕ_h is piecewise linear and $\Gamma_h = \{\phi_h = 0\}$ (Fig. 6.11).

However, $\mathcal{P}_1$ elements for ϕ are not good enough. One should use $\mathcal{P}_2$ instead. Unfortunately, then the level sets of ϕ are no longer $(n-1)$-dimensional simplices but rather more complicated (Fig. 6.12).

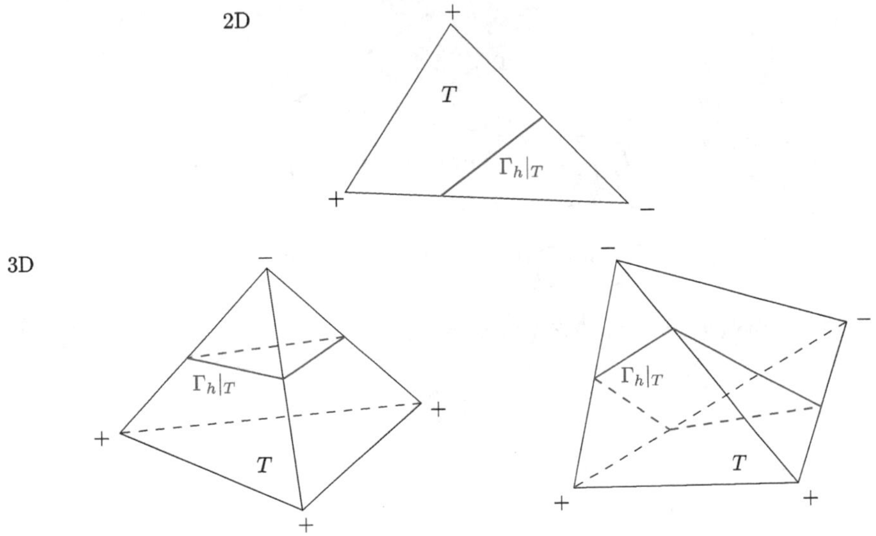

Fig. 6.11 Level set if ϕ_h is piecewise linear

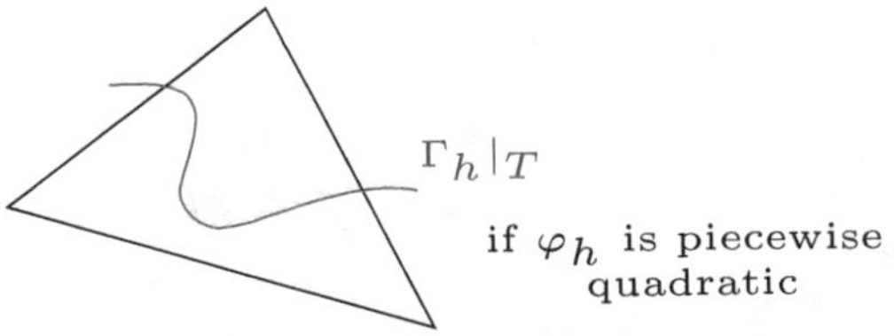

Fig. 6.12 Level set if ϕ_h is of higher order

The remedy is to (virtually) subdivide each element intersected by the level set in 2^n subsimplices and use a piecewise linear interpolation on this sub-complex. Then define the discrete interface Γ_h by

$$\Gamma_h := \{x \in \Omega \mid I_{h/2}(\phi_h)(x) = 0\} \tag{6.15}$$

with $I_{h/2}$ the interpolation operator onto piecewise linears on the sub-complex (Figs. 6.13 and 6.14).

Discontinuous Pressures

We need to do something about the spurious velocities at the interface. Otherwise this could lead to completely wrong computational results if for instance an additional advection-diffusion equation for some chemical species has to be solved.

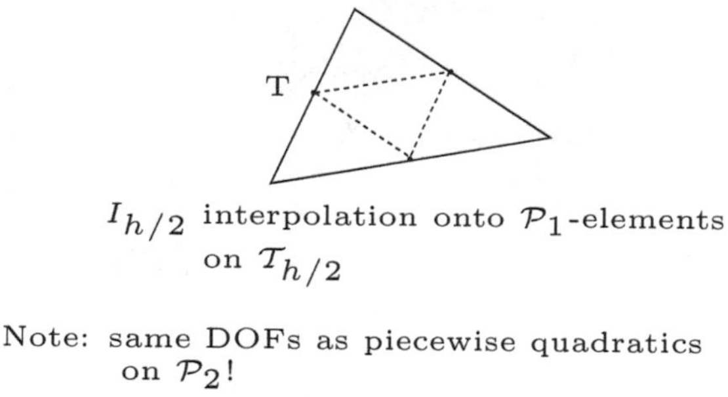

Fig. 6.13 Interpolation operator $I_{h/2}$ onto piecewise linears

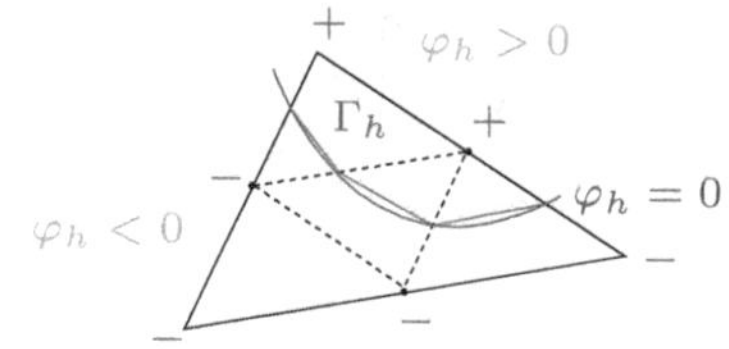

Fig. 6.14 Definition of Γ_h

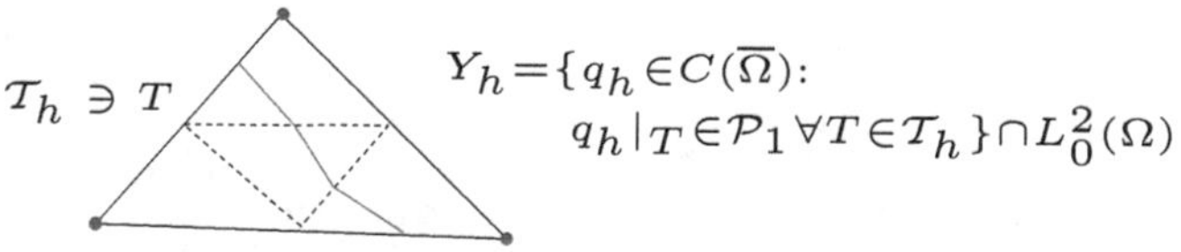

Fig. 6.15 Interface intersecting an element

The situation is as follows. If the discrete pressure functions are continuous across the interface (which is the standard case, Fig. 6.15), then necessarily spurious velocities will occur and the pressure error cannot be better than

$$\|p - p_h\|_{L^2} \sim \sqrt{h}.$$

The idea of *extended finite elements* (XFEM) is now to enrich the discrete pressure space locally by basis functions that are discontinuous across Γ_h, see for instance [74].

For simplicity, let the discrete pressure space Y_h consist of piecewise linear, globally continuous functions. Moreover, let $\psi_1, \dots \psi_M$ denote the Lagrange basis functions of Y_h fulfilling

$$\psi_i(x_j) = \delta_{i,j} \qquad \text{for all vertices } x_j, \quad j = 1, \dots M. \tag{6.16}$$

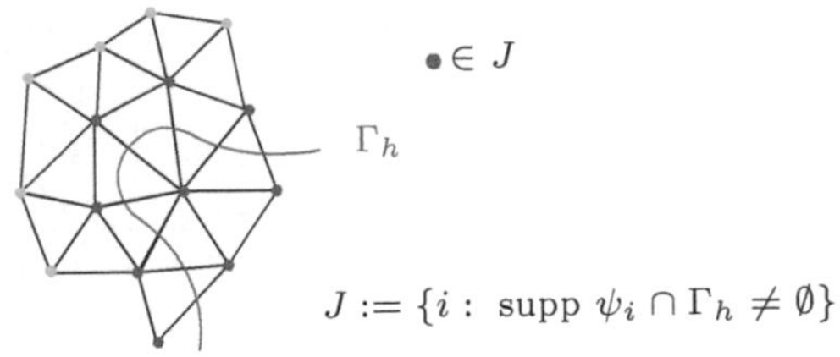

Fig. 6.16 Definition of the index set J

For a given index set $J \subseteq \{1, \dots M\}$ define the XFEM space Y_h^{XFEM} by

$$Y_h^{XFEM} := Y_h \oplus \text{span}\{\psi_j^{XFEM} \mid j \in J\},$$

with functions ψ_j^{XFEM} discontinuous across Γ_h to be determined (Fig. 6.16).

Define

$$H(x) = \tilde{H}(I_{h/2}(\phi_h(x))) = \begin{cases} 0, & x \in \Omega_{1,h}, \\ 1, & x \in \Omega_{2,h} \end{cases}$$

with $\tilde{H}$ the *Heavyside* function. Now for $j \in J$ we define

$$\chi_j(x) := H(x) - H(x_j). \tag{6.17}$$

We note that $\chi_j(x) \in \{-1, 0, 1\}$. With the help of these cut-off functions we define

$$\psi_j^{XFEM}(x) = \psi_j(x)\chi_j(x), \quad \text{for } x \in \Omega.$$

Figures 6.17 and 6.18 show illustrations of these basis functions in one and two dimensions. The implementation of XFEM is not very nice and requires some sophisticated data structure. When Γ_h is moving in time, the space Y_h^{XFEM} has to be built in each time step. Figure 6.19 shows results of simulations that were done without and with XFEM.

Remark 6.4.6

- In certain cases it is enough to add one single degree of freedom to the (discrete) pressure space, see [19].
- Alternative to XFEM:
  In *CutFEM*, one uses non-interface-fitted elements, cuts the standard basis functions at the interface, and adds stabilization terms at the interface, see [35].

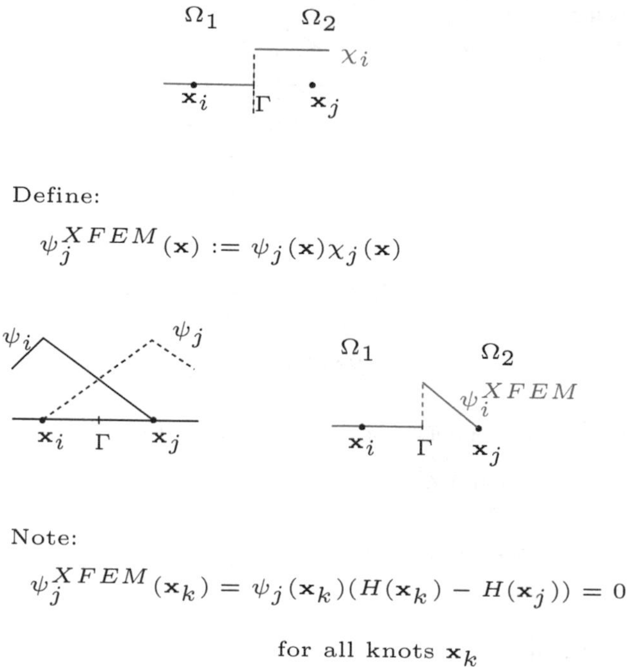
Ω_1
Ω_2
χ_i
$\mathbf{x}_i$
Γ
$\mathbf{x}_j$
Define:
$\psi_j^{XFEM}(\mathbf{x}) := \psi_j(\mathbf{x})\chi_j(\mathbf{x})$
ψ_i
ψ_j
$\mathbf{x}_i$
Γ
$\mathbf{x}_j$
Ω_1
Ω_2
ψ_i^{XFEM}
$\mathbf{x}_i$
Γ
$\mathbf{x}_j$
Note:
$\psi_j^{XFEM}(\mathbf{x}_k) = \psi_j(\mathbf{x}_k)(H(\mathbf{x}_k) - H(\mathbf{x}_j)) = 0$
for all knots $\mathbf{x}_k$

Fig. 6.17 The XFEM basis function ψ_j^{XFEM} in 1d

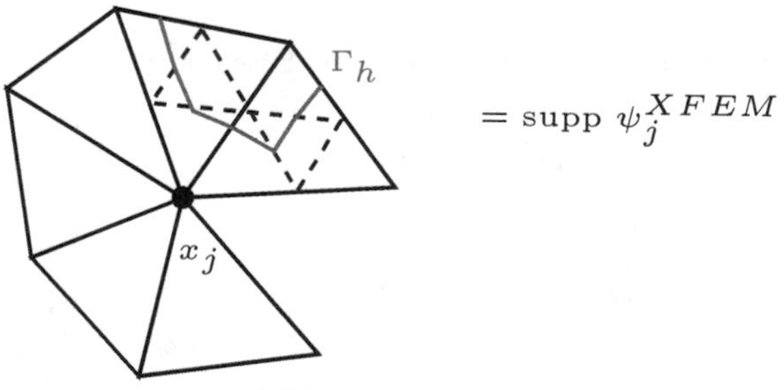
Γ_h
$= \text{supp } \psi_j^{XFEM}$
x_j

Fig. 6.18 The XFEM basis function ψ_j^{XFEM} in 2d

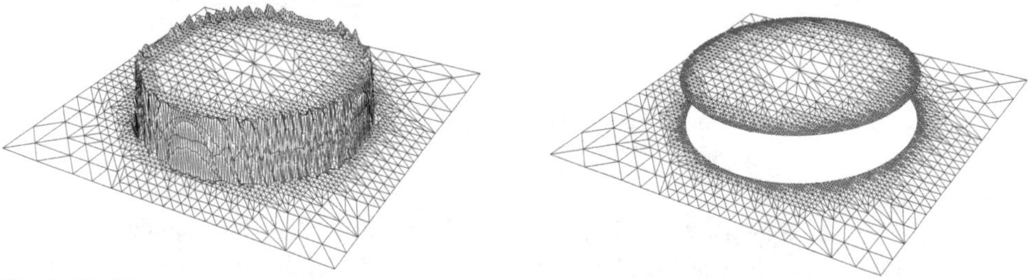

Fig. 6.19 The pressure for the 3d static bubble example. Without (left) and with XFEM (right). Picture taken from [87]

Summarizing, a prototype algorithm for the level-set method (on a fixed grid) with decoupled computation of the flow field and the level-set function reads as follows.

Algorithm 6.4.7 (Level-Set Method for Two Phase Flow) *Let* Γ_h^0, $\Omega_{1,h}^0$, $\Omega_{2,h}^0$, u_h^0 *and* $Y_h^{XFEM,0}$ *be given. Set* $\rho_h^0 = \rho_i$ *and* $\mu_h^0 = \mu_i$ *in* $\Omega_{i,h}^0$, $i = 1, 2$.

In each discrete time step $t_k = k\delta$, $k = 0, 1, 2, \ldots$, *do the following steps.*

1. *Find* $u_h^{k+1} \in X_h$, $p_h^{k+1} \in Y_h^{XFEM,k}$ *fulfilling for all* $\varphi_h \in X_h$ *and* $q_h \in Y_h^{XFEM,k}$

$$
\begin{aligned}
(\rho_h^k \frac{u_h^{k+1} - u_h^k}{\delta}, \varphi_h) + (\rho_h^k (u_h^{k+1}) \cdot \nabla u_h^{k+1}), \varphi_h) - (p_h^{k+1}, \nabla \cdot \varphi_h) \\
+ (2\mu_h^k D(u_h^{k+1}), D(\varphi_h)) + \int_{\Gamma_h^k} \gamma \nabla_{\Gamma_h^k} \cdot \varphi_h d\mathcal{H}^{n-1} = 0,
\end{aligned}
$$

$$
(\nabla \cdot u_h^{k+1}, q_h) = 0.
$$

2. *Find* $\tilde{\phi}_h^{k+1}$ *by performing one time step of the discretized transport equation* (6.14).
3. *Define* $\Gamma_h^{k+1} = \{x \in \Omega \mid I_{h/2}(\tilde{\phi}_h^{k+1})(x) = 0\}$ *(see Eq.* (6.15)*) and*

$$
\Omega_{1,h}^{k+1} = \{x \in \Omega \mid I_{h/2}(\tilde{\phi}_h^{k+1})(x) < 0\}, \; \Omega_{2,h}^{k+1} = \{x \in \Omega \mid I_{h/2}(\tilde{\phi}_h^{k+1})(x) > 0\}.
$$

   Set $\rho_h^{k+1} = \rho_i$ *and* $\mu_h^{k+1} = \mu_i$ *in* $\Omega_{i,h}^{k+1}$, $i = 1, 2$.
4. *Re-initialize* $\tilde{\phi}_h^{k+1}$ *to a discrete (nearly) signed-distance function* ϕ_h^{k+1}.
5. *Construct* $Y_h^{XFEM,k+1}$.

7 Exercises

Exercise 7.1 Prove Identity (2.7).

Exercise 7.2 Prove the assertion in Proposition 2.3.3:

Let ϕ with $\nabla\phi(x) \neq 0$ for $x \in \Gamma$ and Γ be smooth and given by $\Gamma = \{x \in \Omega \mid \phi(x) = 0\}$ be the zero level set of ϕ. Let the unit normal field ν on $V \cap \Gamma$ be given by $\nu = \tilde{\nu}_{|V\cap\Gamma}$, with $\tilde{\nu} : V \to I\!R^n$ defined as $\tilde{\nu}(y) = \dfrac{\nabla\phi(y)}{|\nabla\phi(y)|}$. Prove that

$$\kappa(x) = -\nabla \cdot \frac{\nabla\phi(x)}{|\nabla\phi(x)|}, \qquad \text{for } x \in \Gamma.$$

Exercise 7.3 Let $\Gamma = (\Gamma(t))_{t\in(0,T)}$ be a family of evolving hypersurfaces in $I\!R^n$. Show that a curve γ with the properties required in Definition 2.7.1 (i) exists and that the definition of the normal velocity V is independent of the choice of γ.

Hint: Describe Γ locally around (t_0, x_0) as a level set of a function $\phi : I\!R \times I\!R^n \to I\!R$.

Exercise 7.4 Let Γ be an evolving hypersurface and $(t_0, x_0) \in \Gamma$ with unit normal $\nu(t_0, x_0)$ and normal velocity $V(t_0, x_0)$. Show that there exists a curve $\gamma : (t_0-\delta, t_0+\delta) \to \mathbb{R}^n$ satisfying $\gamma(t) \in \Gamma(t)$, $\gamma(t_0) = x_0$ and $\frac{d\gamma}{dt}(t_0) = V(t_0, x_0)\nu(t_0, x_0)$.

Exercise 7.5 Let Γ be an orientable C^3-hypersurface with normal vector field ν and for a function $f : \Gamma \to I\!R$ set $\partial_{s_i} f := (\nabla_\Gamma f)_i$.

Show that the commutator rule

$$\partial_{s_j}\partial_{s_i} f = \partial_{s_i}\partial_{s_j} f + [(\nabla_\Gamma \nu)\nabla_\Gamma f]_i \nu_j - [(\nabla_\Gamma \nu)\nabla_\Gamma f]_j \nu_i \tag{7.1}$$

holds.

E. Bänsch et al., *Interfaces: Modeling, Analysis, Numerics*,
Oberwolfach Seminars 51, https://doi.org/10.1007/978-3-031-35550-9_7

Exercise 7.6 Let Γ be an orientable C^3-hypersurface with normal vector field ν. Then it holds that

$$\nabla_\Gamma \kappa = -\Delta_\Gamma \nu - |\nabla_\Gamma \nu|^2 \nu \qquad \text{on } \Gamma. \tag{7.2}$$

Here $|A| = \sqrt{tr(A^T A)}$ is the Frobenius norm of a matrix A.

Readers with less experience in geometry can show this for curves in the plane. For more experienced readers who want to deal with the general case we have the following hint.

Hint: Consider $\Delta_\Gamma \nu_j = \sum_{i=1}^n \partial_{s_i} \partial_{s_i} \nu_j$, use the symmetry of the Weingarten map, use the commutator rule (7.1), and the fact that $\kappa = -\nabla_\Gamma \cdot \nu$.

Exercise 7.7 Let Γ be an orientable C^3 evolving hypersurface with normal vector field ν.

(i) Prove the identity

$$D_t \kappa = \Delta_\Gamma V + V|\nabla_\Gamma \nu|^2 + v_\tau \cdot \nabla_\Gamma \kappa. \tag{7.3}$$

(ii) Show that the normal time derivative of the mean curvature satisfies

$$\partial_t^\square \kappa = \Delta_\Gamma V + V|\nabla_\Gamma \nu|^2. \tag{7.4}$$

Here $|A| = \sqrt{tr(A^T A)}$ is the Frobenius norm of a matrix A.

Readers with less experience in geometry show this for curves in the plane. For more experienced readers who want to deal with the general case we have the following hint.

Hint: Derive the following rule for the commutation of D_t and the divergence of the normal field $\nu : \Gamma \to I\!R^n$

$$D_t(\nabla_\Gamma \cdot \nu) = \nabla_\Gamma \cdot (D_t \nu) - V|\nabla_\Gamma \nu|^2 - \nabla_\Gamma v_\tau : \nabla_\Gamma \nu$$

and then use the results of Exercise 7.6.

Exercise 7.8 Prove the Transport Theorem 3.6.1.

Exercise 7.9 Let $\Omega \subset I\!R^n$ be a domain and $\theta : (0, T) \times \Omega \to I\!R$ a continuous distributional solution of

$$\partial_t \big(\theta + \chi_{\{\theta > \theta_M\}}\big) = \Delta\theta\,,$$

cf. (3.29). Let us assume that the set $\Gamma = \{(t,x) \mid \theta(t,x) = \theta_M\}$ is a smooth evolving hypersurface with $\Gamma(t) \subset\subset \Omega$ for all $t \in (0,T)$. We now set $\Omega_s(t) = \{x \in \Omega \mid \theta < \theta_M\}$, $\Omega_\ell(t) = \{x \in \Omega \mid \theta > \theta_M\}$, $Q_s = \{(t,x) \in (0,T) \times \Omega \mid x \in \Omega_s(t)\}$ and $Q_l = \{(t,x) \in (0,T) \times \Omega \mid x \in \Omega_l(t)\}$. We assume that $\theta_{|Q_s}$ and $\theta_{|Q_\ell}$ individually have twice continuously differentiable extensions onto $\Gamma(t)$.

Show that under these assumptions

$$V = [-\nabla\theta]_s^\ell \cdot n \quad \text{on} \quad \Gamma(t)$$

holds.

Exercise 7.10 Show that Eqs. (4.20) and (4.21) are equivalent.

Exercise 7.11 Show that

$$|\Gamma_h^{m+1}| \leq |\Gamma_h^m|$$

holds for the fully discrete scheme for mean curvature flow introduced in Sect. 4.3.4.

Hint: Use Lemma 1 in Section 3 of [10] or Lemma 55 of [23].

Exercise 7.12 Consider a time dependent smooth curve $f : [0,T) \times I \to \mathbb{R}^n$, $f = f(t,x)$ and put

$$\partial_t f = V + \varphi\tau$$

where V is the normal velocity, $\tau = \frac{\partial_x f}{|\partial_x f|} = \partial_s f$ the unit tangent, and $\varphi = \langle \partial_t f, \tau\rangle$. Let $k = \partial_s \tau$ be the curvature vector. Show that

$$\begin{aligned}
\partial_t \partial_s &= \partial_s \partial_t + (\langle k, V\rangle - \partial_s\varphi)\partial_s\,,\\
\partial_t \tau &= \nabla_s V + \varphi k\,,\\
\nabla_t k &= \nabla_s^2 V + \langle k, V\rangle k + \varphi \nabla_s k
\end{aligned}$$

hold, where ∇_s is as in (4.42) and $\nabla_t \phi = \partial_t \phi - \langle \partial_t \phi, \tau\rangle\tau$ for any $\phi : [0,T) \times I \to \mathbb{R}^n$.

Exercise 7.13 Derive the first variation (4.41) for the elastic energy $E(f) = \frac{1}{2}\int_I |k|^2 ds$, i.e. show that

$$\frac{d}{d\varepsilon}\Big|_{\varepsilon=0} E(f_\varepsilon) = \int_I \langle \nabla_s^2 k + \frac{1}{2}|k|^2 k, \phi\rangle ds$$

holds for smooth variations $f_\epsilon(x) = f(x) + \epsilon\phi(x)$, $\phi : I \to \mathbb{R}^n$, and with ∇_s defined as in (4.42).

Exercise 7.14 Show the existence of a constant $c > 0$ such that inequality (4.49) holds for any smooth closed curve f.

Hint: First show that if $u : I \to \mathbb{R}$, $u = u(x)$, is a smooth map along the curve $f : I \to \mathbb{R}^n$, $f = f(x)$, such that

$$\int_I u\,ds = \int_I u(x)|f_x(x)|dx = 0$$

then

$$\int_I u^2 ds \leq \mathscr{L}(f)^2 \int_I (u_s)^2 ds$$

holds, where $\partial_s = \frac{1}{|f_x|}\partial_x$, $ds = |f_x|dx$, and $\mathscr{L}(f) = \int_I ds = \int_I |f_x|dx$ is the length of the curve.

Exercise 7.15 Verify Eq. (4.62) for (f_h, k_h) being a solution to (4.60) and (4.61).

Exercise 7.16 Show

$$\int_\Omega |\nabla \chi_E| = \mathscr{H}^{n-1}(\partial E)$$

if $E \subset\subset \Omega$, $E, \Omega \subset I\!R^n$ open, with E having a smooth boundary.

Exercise 7.17 Let the assumptions of Theorem 5.8.1 hold and assume in addition that there exists a constant $c > 0$ such that for all $\varphi_1, \varphi_2 \in I\!R$ it holds

$$(\psi'(\varphi_1) - \psi'(\varphi_2))(\varphi_1 - \varphi_2) \geq -c|\varphi_1 - \varphi_2|^2.$$

Show that under these assumptions and under the regularity properties stated, the solution in Theorem 5.8.1 is unique.

Hint: Use the H^{-1}-gradient flow property.

References

1. H. Abels, On a diffuse interface model for two-phase flows of viscous, incompressible fluids with matched densities. Arch. Ration. Mech. Anal. **194**(2), 463–506 (2009)
2. H. Abels, M. Wilke, Convergence to equilibrium for the Cahn-Hilliard equation with a logarithmic free energy. Nonlinear Anal. **67**(11), 3176–3193 (2007)
3. H. Abels, H. Garcke, G. Grün, Thermodynamically consistent, frame indifferent diffuse interface models for incompressible two-phase flows with different densities. Math. Models Methods Appl. Sci. **22**(3), 1150013 (2012)
4. H. Abels, D. Depner, H. Garcke, Existence of weak solutions for a diffuse interface model for two-phase flows of incompressible fluids with different densities. J. Math. Fluid Mech. **15**(3), 453–480 (2013)
5. F. Almgren, J.E. Taylor, L. Wang, Curvature-driven flows: a variational approach. SIAM J. Control Optim. **31**(2), 387–438 (1993)
6. H.W. Alt, *Linear Functional Analysis*. Universitext (Springer, London, 2016). An application-oriented introduction
7. L. Ambrosio, N. Fusco, D. Pallara, *Functions of Bounded Variation and Free Discontinuity Problems*. Oxford Mathematical Monographs (Clarendon Press, Oxford, 2000)
8. L. Ambrosio, N. Gigli, G. Savaré, *Gradient Flows in Metric Spaces and in the Space of Probability Measures*. Lectures in Mathematics ETH Zürich, 2nd edn. (Birkhäuser Verlag, Basel, 2008)
9. R.F. Ausas, G.C. Buscaglia, S.R. Idelsohn, A new enrichment space for the treatment of discontinuous pressures in multi-fluid flows. Int. J. Numer. Methods Fluids **70**(7), 829–850 (2012)
10. E. Bänsch, Finite element discretization of the Navier–Stokes equations with a free capillary surface. Numer. Math. **88**(2), 203–235 (2001)
11. E. Bänsch, A. Schmidt, Free boundary problems in fluids and materials, in *Handbook of Numerical Analysis*, vol. 21 (Elsevier, Amsterdam, 2020), pp. 555–619
12. E. Bänsch, S. Basting, R. Krahl, Numerical simulation of two-phase flows with heat and mass transfer. Discrete Contin. Dyn. Syst. **35**(6), 2325–2347 (2015)
13. J.W. Barrett, J.F. Blowey, H. Garcke, Finite element approximation of the Cahn-Hilliard equation with degenerate mobility. SIAM J. Numer. Anal. **37**(1), 286–318 (1999)
14. J.W. Barrett, H. Garcke, R. Nürnberg, On the variational approximation of combined second and fourth order geometric evolution equations. SIAM J. Sci. Comput. **29**(3), 1006–1041 (2007)
15. J.W. Barrett, H. Garcke, R. Nürnberg, A parametric finite element method for fourth order geometric evolution equations. J. Comput. Phys. **222**(1), 441–462 (2007)

E. Bänsch et al., *Interfaces: Modeling, Analysis, Numerics*,
Oberwolfach Seminars 51, https://doi.org/10.1007/978-3-031-35550-9

16. J.W. Barrett, H. Garcke, R. Nürnberg, On the parametric finite element approximation of evolving hypersurfaces in $\mathbb{R}^3$. J. Comput. Phys. **227**(9), 4281–4307 (2008)
17. J.W. Barrett, H. Garcke, R. Nürnberg, On stable parametric finite element methods for the Stefan problem and the Mullins–Sekerka problem with applications to dendritic growth. J. Comput. Phys. **229**(18), 6270–6299 (2010)
18. J.W. Barrett, H. Garcke, R. Nürnberg, Numerical computations of facetted pattern formation in snow crystal growth. Phys. Rev. E **86**(1), 011604 (2012)
19. J.W. Barrett, H. Garcke, R. Nürnberg, Eliminating spurious velocities with a stable approximation of viscous incompressible two-phase Stokes flow. Comput. Methods Appl. Mech. Eng. **267**, 511–530 (2013)
20. J.W. Barrett, H. Garcke, R. Nürnberg, On the stable discretization of strongly anisotropic phase field models with applications to crystal growth. ZAMM Z. Angew. Math. Mech. **93**(10–11), 719–732 (2013)
21. J.W. Barrett, H. Garcke, R. Nürnberg, On the stable numerical approximation of two-phase flow with insoluble surfactant. ESAIM Math. Model. Numer. Anal. **49**(2), 421–458 (2015)
22. J.W. Barrett, H. Garcke, R. Nürnberg, Stable finite element approximations of two-phase flow with soluble surfactant. J. Comput. Phys. **297**, 530–564 (2015)
23. J.W. Barrett, H. Garcke, R. Nürnberg, Parametric finite element approximations of curvature-driven interface evolutions, in *Handbook of Numerical Analysis*, vol. 21 (Elsevier, Amsterdam, 2020), pp. 275–423
24. G. Bellettini, M. Paolini, Anisotropic motion by mean curvature in the context of Finsler geometry. Hokkaido Math. J. **25**(3), 537–566 (1996)
25. H. Benninghoff, H. Garcke, Efficient image segmentation and restoration using parametric curve evolution with junctions and topology changes. SIAM J. Imag. Sci. **7**(3), 1451–1483 (2014)
26. H. Benninghoff, H. Garcke, Segmentation of three-dimensional images with parametric active surfaces and topology changes. J. Sci. Comput. **72**(3), 1333–1367 (2017)
27. J.F. Blowey, C.M. Elliott, The Cahn-Hilliard gradient theory for phase separation with nonsmooth free energy. Eur. J. Appl. Math. **2**(3), 233–280 (1991)
28. J. Bosch, M. Stoll, Preconditioning for vector-valued Cahn-Hilliard equations. SIAM J. Sci. Comput. **37**(5), S216–S243 (2015)
29. F. Boyer, P. Fabrie, *Mathematical Tools for the Study of the Incompressible Navier-Stokes Equations and Related Models*. Applied Mathematical Sciences, vol. 183 (Springer, New York, 2013)
30. J.U. Brackbill, D. Kothe, C. Zemach, A continuum method for modeling surface tension. J. Comput. Phys. **100**(2), 335–354 (1992)
31. A. Braides, *Γ-Convergence for Beginners*. Oxford Lecture Series in Mathematics and its Applications, vol. 22 (Oxford University Press, London, 2005), xii+217pp.
32. K.A. Brakke, *The Motion of a Surface by its Mean Curvature*. Mathematical Notes, vol. 20 (Princeton University Press, Princeton, 1978), i+252pp.
33. L. Bronsard, F. Reitich, On three-phase boundary motion and the singular limit of a vector-valued ginzburg-landau equation. Arch. Rational Mech. Anal. **124**(4), 355–379 (1993)
34. A.N. Brooks, T.J.R. Hughes, Streamline upwind/Petrov-Galerkin formulations for convection dominated flows with particular emphasis on the incompressible Navier-Stokes equations. Comput. Methods Appl. Mech. Eng. **32**(1–3), 199–259 (1982)
35. E. Burman, S. Claus, P. Hansbo, M.G. Larson, A. Massing, Cutfem: discretizing geometry and partial differential equations. Int. J. Numer. Methods Eng. **104**, 472–501 (2015)
36. G.C. Buscaglia, R.F. Ausas, Variational formulations for surface tension, capillarity and wetting. Comput. Methods Appl. Mech. Eng. **200**, 3011–3025 (2011)

37. Y.G. Chen, Y. Giga, S. Goto, Uniqueness and existence of viscosity solutions of generalized mean curvature flow equations. J. Differ. Geom. **33**(3), 749–786 (1991)
38. S. Claus, P. Kerfriden, A CutFEM method for two-phase flow problems. Comput. Methods Appl. Mech. Eng. **348**, 185–206 (2019)
39. R.G. Cox, The dynamics of the spreading of liquids on a solid surface. Part 1. Viscous flow. J. Fluid Mech. **168**, 169–194 (1986)
40. G. Dal Maso, *An Introduction to* Γ*-Convergence*, vol. 8 (Birkhäuser, Boston, 1993)
41. T.A. Davis, Algorithm 832: UMFPACK V4.3—an unsymmetric-pattern multifrontal method. ACM Trans. Math. Softw. **30**(2), 196–199 (2004)
42. K. Deckelnick, G. Dziuk, On the approximation of the curve shortening flow, in *Calculus of Variations, Applications and Computations (Pont-à-Mousson, 1994)*, ed. by C. Bandle, J. Bemelmans, M. Chipot, J. Saint Jean Paulin. Pitman Research Notes in Mathematics Series (Longman Scientific & Technical, Harlow, 1994), pp. 100–108
43. K. Deckelnick, G. Dziuk, Convergence of numerical schemes for the approximation of level set solutions to mean curvature flow, in *Numerical Methods for Viscosity Solutions and Applications*, ed. by M. Falcone, C. Makridakis. Advanced Mathematics and Applications Sciences, vol. 59 (World Scientific, Singapore, 2001), pp. 77–94
44. K. Deckelnick, G. Dziuk, A fully discrete numerical scheme for weighted mean curvature flow. Numer. Math. **91**(3), 423–452 (2002)
45. K. Deckelnick, G. Dziuk, Error analysis for the elastic flow of parametrized curves. Math. Comput. **78**(266), 645–671 (2009)
46. K. Deckelnick, G. Dziuk, C.M. Elliott, Computation of geometric partial differential equations and mean curvature flow. Acta Numer. **14**, 139–232 (2005)
47. D. Depner, H. Garcke, Y. Kohsaka, Mean curvature flow with triple junctions in higher space dimensions. Arch. Ration. Mech. Anal. **211**(1), 301–334 (2014)
48. M.P. do Carmo, *Riemannian Geometry*. Mathematics: Theory & Applications (Birkhhäuser, Boston, 1992). Translated from the second Portuguese edition by Francis Flaherty
49. J. Donea, S. Giuliani, J.P. Halleux, An arbitrary Lagrangian–Eulerian finite element method for transient dynamic fluid–structure interactions. Comput. Methods App. Mech. Eng. **33**, 689–723 (1982)
50. Q. Du, X. Feng, The phase field method for geometric moving interfaces and their numerical approximations, in *Geometric Partial Differential Equations. Part I.* Handbook of Numerical Analysis, vol. 21 (Elsevier/North-Holland, Amsterdam, 2020), pp. 425–508
51. G. Dziuk, An algorithm for evolutionary surfaces. Numer. Math. **58**(1), 603–611 (1990)
52. G. Dziuk, Convergence of a semi-discrete scheme for the curve shortening flow. Math. Models Methods Appl. Sci. **4**(4), 589–606 (1994)
53. G. Dziuk, C.M. Elliott, Finite elements on evolving surfaces. IMA J. Numer. Anal. **27**(2), 262–292 (2007)
54. G. Dziuk, C.M. Elliott, Finite element methods for surface PDEs. Acta Numer. **22**, 289–396 (2013)
55. G. Dziuk, E. Kuwert, R. Schätzle, Evolution of elastic curves in $\mathbb{R}^n$: existence and computation. SIAM J. Math. Anal. **33**(5), 1228–1245 (2002)
56. C. Eck, M. Fontelos, G. Grün, F. Klingbeil, O. Vantzos, On a phase-field model for electrowetting. Interfaces Free Bound. **11**(2), 259–290 (2009)
57. C. Eck, H. Garcke, P. Knabner, *Mathematical Modeling* (Springer, Berlin, 2017)
58. K. Ecker, *Regularity Theory for Mean Curvature Flow*. Progress in Nonlinear Differential Equations and Their Applications, vol. 57, xiv+165pp. (Birkhäuser, Boston, 2004)
59. K. Ecker, Heat equations in geometry and topology. Jahresber. Deutsch. Math.-Verein. **110**(3), 117–141 (2008)

60. C.M. Elliott, *The Cahn–Hilliard Model for the Kinetics of Phase Separation*. International Series of Numerical Mathematics, vol. 88 (Birkhäuser Verlag, Basel, 1989), pp. 35–73
61. C.M. Elliott, H. Fritz, On approximations of the curve shortening flow and of the mean curvature flow based on the DeTurck trick. IMA J. Numer. Anal. **37**(2), 543–603 (2017)
62. C.M. Elliott, H. Garcke, On the Cahn-Hilliard equation with degenerate mobility. SIAM J. Math. Anal. **27**(2), 404–423 (1996)
63. C.M. Elliott, D.A. French, F.A. Milner, A second order splitting method for the Cahn-Hilliard equation. Numer. Math. **54**(5), 575–590 (1989)
64. J. Escher, G. Simonett, A center manifold analysis for the Mullins-Sekerka model. J. Differ. Equ. **143**(2), 267–292 (1998)
65. P. Esser, J. Grande, An accurate and robust finite element level set redistancing method. IMA J. Numer. Anal. **35**(4), 1913–1933 (2015)
66. L.C. Evans, *Partial Differential Equations*. Graduate Studies in Mathematics, vol. 19, 2nd edn. (American Mathematical Society, Providence, 2010)
67. L.C. Evans, J. Spruck, Motion of level sets by mean curvature. I. J. Differ. Geom. **33**(3), 635–681 (1991)
68. P.C. Fife, *Dynamics of Internal Layers and Diffusive Interfaces*. CBMS-NSF Regional Conference Series in Applied Mathematics, vol. 53 (Society for Industrial and Applied Mathematics (SIAM), Philadelphia, 1988)
69. P.C. Fife, *Barrett Lecture Notes* (University of Tennessee, Knoxville, 1991)
70. P.C. Fife, *Models for Phase Separation and Their Mathematics* (KTK, Tokyo, 1993)
71. P.C. Fife, O. Penrose, Interfacial dynamics for thermodynamically consistent phase-field models with nonconserved order parameter. Electron. J. Differ. Equ. **16**, 1–49 (1995)
72. M. Fried, Berechnung des Krümmungsflusses von Niveauflächen. Diploma Thesis, Institut für Angewandte Mathematik, Universität Freiburg (1993)
73. M. Fried, A level set based finite element algorithm for the simulation of dendritic growth. Comput. Visual. Sci. **7**(2), 97–110 (2004)
74. T.P. Fries, T. Belytschko, The extended/generalized finite element method: an overview of the method and its applications. Int. J. Numer. Methods Eng. **84**, 253–304 (2010)
75. M. Gage, R.S. Hamilton, The heat equation shrinking convex plane curves. J. Differ. Geom. **23**, 69–96 (1986)
76. H. Garcke, Curvature driven interface evolution. Jahresbericht Dtsch. Math.–Ver. **115**(2), 63–100 (2013)
77. H. Garcke, B. Stinner, Second order phase field asymptotics for multi-component systems. Interfaces Free Bound. **8**(2), 131–157 (2006)
78. H. Garcke, B. Nestler, B. Stinner, A diffuse interface model for alloys with multiple components and phases. SIAM J. Appl. Math. **64**(3), 775–799 (2004)
79. H. Garcke, M. Hinze, C. Kahle, A stable and linear time discretization for a thermodynamically consistent model for two-phase incompressible flow. Appl. Numer. Math. **99**, 151–171 (2016)
80. H. Garcke, R. Nürnberg, Q. Zhao, Structure-preserving discretizations of two-phase Navier-Stokes flow using fitted and unfitted approaches. J. Comput. Phys. **489**, Paper No. 112276 (2023)
81. F. Gibou, L. Chen, D. Nguyen, S. Banerjee, A level set based sharp interface method for the multiphase incompressible Navier–Stokes equations with phase change. J. Comput. Phys. **222**(2), 536–555 (2007)
82. Y. Giga, *Surface Evolution Equations. A Level Set Approach*. Monographs in Mathematics, vol. 99 (Birkhäuser Verlag, Basel, 2006)
83. D. Gilbarg, N.S. Trudinger, *Elliptic Partial Differential Equations of Second Order*. Classics in Mathematics (Springer, Berlin, 2001). Reprint of the 1998 edition

84. E. Giusti, *Minimal Surfaces and Functions of Bounded Variation*. Monographs in Mathematics, vol. 80 (Birkhäuser Verlag, Basel, 1984)
85. C. Gräser, R. Kornhuber, U. Sack, Time discretizations of anisotropic Allen-Cahn equations. IMA J. Numer. Anal. **33**(4), 1226–1244 (2013)
86. M. Grayson, The heat equation shrinks embedded plane curves to round points. J. Differ. Geom. **26**, 285–314 (1987)
87. S. Groß, A. Reusken, *Numerical Methods for Two-Phase Incompressible Flows*. Springer Series in Computational Mathematics, vol. 40 (Springer, Berlin, 2011)
88. M.E. Gurtin, *An Introduction to Continuum Mechanics* (Academic Press, Cambridge, 1981)
89. M.E. Gurtin, D. Polignone, J. Viñals, Two-phase binary fluids and immiscible fluids described by an order parameter. Math. Models Methods Appl. Sci. **6**(6), 815–831 (1996)
90. C.W. Hirt, B.D. Nichols, Volume of fluid (VOF) method for the dynamics of free boundaries. J. Comput. Phys. **39**(1), 201–225 (1981)
91. T.J.R. Hughes, W. Liu, T.K. Zimmermann, Lagrangian-Eulerian finite element formulation for incompressible viscous flows. Comput. Methods Appl. Mech. Eng. **29**(3), 329–349 (1981)
92. G. Huisken, Flow by mean curvature of convex surfaces into spheres. J. Differ. Geom. **20**(1), 237–266 (1984)
93. G. Huisken, A. Polden, *Geometric Evolution Equations for Hypersurfaces*. Lecture Notes in Mathematics, vol. 1713 (Springer, Berlin, 1999).
94. T. Ilmanen, *Elliptic Regularization and Partial Regularity for Motion by Mean Curvature*. Memoirs of the American Mathematical Society, vol. 108(520), x+90pp. (American Mathematical Society, Providence, 1994)
95. K. Ishimi, H. Hikita, M.N. Esmail, Dynamic contact angles on moving plates. AIChE **32**, 486–492 (1986)
96. D. Jacqmin, Calculation of two-phase Navier–Stokes flows using phase-field modeling. J. Comput. Phys. **155**(1), 96–127 (1999)
97. D. Jamet, O. Lebaigue, N. Coutris, J.M. Delhaye, The second gradient method for the direct numerical simulation of liquid–vapor flows with phase change. J. Comput. Phys. **169**(2), 624–651 (2001)
98. T.S. Jiang, S.G. OH, J.C. Slattery, Correlation for dynamic contact angle. J. Colloid Interface Sci. **69**, 74–77 (1979)
99. V. John, E. Schmeyer, Finite element methods for time-dependent convection-diffusion-reaction equations with small diffusion. Comput. Methods Appl. Mech. Eng. **198**(3–4), 475–494 (2008)
100. J. Jost, *Partial Differential Equations*. Graduate Texts in Mathematics, vol. 214 (Springer New York, 2002). Translated and revised from the 1998 German original by the author
101. C.E. Kees, I. Akkerman, M.W. Farthing, Y. Bazilevs, A conservative level set method suitable for variable-order approximations and unstructured meshes. J. Comput. Phys. **230**(12), 4536–4558 (2011)
102. P. Knabner, L. Angermann, *Numerical Methods for Elliptic and Parabolic Partial Differential Equations*. Texts in Applied Mathematics, vol. 44 (Springer, Cham, 2021). ©2021. With contributions by Andreas Rupp, Second extended edition [of 1988268]
103. B. Kovács, Numerical surgery for mean curvature flow of surfaces (2022, arXiv-Preprint). https://arxiv.org/abs/2210.14046
104. B. Kovacs, B. Li, C. Lubich, A convergent evolving finite element algorithm for mean curvature flow of closed surfaces. Numer. Math. **143**(4), 797–853 (2019)
105. W. Kühnel, *Differential Geometry*. Student Mathematical Library, vol. 77 (American Mathematical Society, Providence, 2015). Curves—surfaces—manifolds, Third edition [of MR1882174], Translated from the 2013 German edition by Bruce Hunt, with corrections and additions by the author

106. O.A. Ladyženskaja, V.A. Solonnikov, Ural'ceva, in *Linear and Quasi-linear Equations of Parabolic Type*. Transactions of Mathematical Monographs, vol. 23 (American Mathematical Society, Providence, 1998)
107. R.J. LeVeque, Z. Li, Immersed interface methods for Stokes flow with elastic boundaries or surface tension. SIAM J. Sci. Comput. **18**(3), 709–735 (1997)
108. B. Li, Convergence of Dziuk's semidiscrete finite element method for mean curvature flow of closed surfaces with high-order finite elements. SIAM J. Numer. Anal. **59**(3), 1592–1617 (2021)
109. K.G. Libbrecht, *Morphogenesis on Ice: The Physics of Snow Crystals*. Number 1 in Engineering & Science (2001)
110. G. Lieberman, *Second Order Parabolic Differential Equations* (World Scientific, Singapore, 2005)
111. I.S. Liu, *Continuum Mechanics. Advanced Texts in Physics* (Springer, Berlin, 2002)
112. J. Lowengrub, L. Truskinovsky, Quasi-incompressible Cahn-Hilliard fluids and topological transitions. R. Soc. Lond. Proc. Ser. A Math. Phys. Eng. Sci. **454**, 2617–2654 (1998)
113. S. Luckhaus, Solutions for the two-phase Stefan problem with the Gibbs–Thomson law for the melting temperature. Eur. J. Appl. Math. **1**(2), 101–111 (1990)
114. S. Luckhaus, T. Sturzenhecker, Implicit time discretization for the mean curvature flow equation. Calc. Var. Partial Differ. Equ. **3**(2), 253–271 (1995)
115. A. Lunardi, *Analytic Semigroups and Optimal Regularity in Parabolic Problems* (Birkhäuser, Basel, 1995)
116. C. Mantegazza, *Lecture Notes on Mean Curvature Flow*. Progress in Mathematics, vol. 290 (Birkhäuser/Springer, Basel, 2011), xii+166pp.
117. C. Mantegazza, A. Pluda, M. Pozzetta, A survey of the elastic flow of curves and networks. Milan J. Math. **89**(1), 59–121 (2021)
118. K. Mikula, D. Ševčovič, Evolution of plane curves driven by a nonlinear function of curvature and anisotropy. SIAM J. Appl. Math. **61**(5), 1473–1501 (2001)
119. A. Miranville, *The Cahn-Hilliard Equation*. CBMS-NSF Regional Conference Series in Applied Mathematics, vol. 95 (Society for Industrial and Applied Mathematics, Philadelphia, 2019). Recent advances and applications
120. L. Modica, The gradient theory of phase transitions and the minimal interface criterion. Arch. Rational Mech. Anal. **98**, 123–142 (1987)
121. L. Modica, S. Mortola, Un esempio di γ-convergenza. Boll. Un. Mat. Ital. B **14**(5), 285–299 (1977)
122. A. Novick-Cohen, The Cahn–Hilliard equation: mathematical and modeling perspectives. Adv. Math. Sci. Appl. **8**(2), 965–985 (1998)
123. S. Osher, R.P. Fedkiw, Level set methods: an overview and some recent results. J. Comput. Phys. **169**, 463–502 (2001)
124. S. Osher, J.A. Sethian, Fronts propagating with curvature-dependent speed: algorithms based on Hamilton–Jacobi formulations. J. Comput. Phys. **79**(1), 12–49 (1988)
125. O. Penrose, P.C. Fife, Thermodynamically consistent models of phase-field type for the kinetics of phase transitions. Phys. D **43**(1), 44–62 (1990)
126. C.S. Peskin, The immersed boundary method. Acta Numer. **11**, 479–517 (2002)
127. L.M. Pismen, Y. Pomeau, Disjoining potential and spreading of thin liquid layers in the diffuse-interface model coupled to hydrodynamics. Phys. Rev. E **62**(2), 2480–2492 (2000)
128. P. Pozzi, Anisotropic curve shortening flow in higher codimension. Math. Methods Appl. Sci. **30**(11), 1243–1281 (2007)
129. J. Prüss, G. Simonett, *Moving Interfaces and Quasilinear Parabolic Evolution Equations*. Monographs in Mathematics, vol. 105 (Springer, Berlin, 2016)

130. M. Quezada de Luna, D. Kuzmin, C.E. Kees, A monolithic conservative level set method with built-in redistancing. J. Comput. Phys. **379**, 262–278 (2019)
131. A. Reusken, A finite element level set redistancing method based on gradient recovery. SIAM J. Numer. Anal. **51**(5), 2723–2745 (2013)
132. A. Reusken, P. Esser, Analysis of time discretization methods for Stokes equations with a nonsmooth forcing term. Numer. Math. **126**(2), 293–319 (2014)
133. M. Rumpf, A variational approach to optimal meshes. Numer. Math. **72**(4), 523–540 (1996)
134. G. Russo, P. Smereka, A remark on computing distance functions. J. Comput. Phys. **163**(1), 51–67 (2000)
135. R.I. Saye, J.A. Sethian, A review of level set methods to model interfaces moving under complex physics: recent challenges and advances, in *Geometric Partial Differential Equations. Part I*. Handbook of Numerical Analysis, vol. 21 (Elsevier/North-Holland, Amsterdam, 2020), pp. 509–554. ©2020
136. R. Scardovelli, S. Zaleski, Direct numerical simulation of free-surface and interfacial flow. Ann. Rev. Fluid Mech. **31**, 567–603 (1999)
137. J. Schlottke, B. Weigand, Direct numerical simulation of evaporating droplets. J. Comput. Phys. **227**(10), 5215–5237 (2008)
138. L.E. Scriven, Dynamics of a fluid interface equation of motion for Newtonian surface fluids. Chem. Eng. Sci. **12**(2), 98–108 (1960)
139. J.A. Sethian, *Level Set Methods and Fast Marching Methods*. Cambridge Monographs on Applied and Computational Mathematics, vol. 3 (Cambridge University Press, Cambridge, 1999). Evolving interfaces in computational geometry, fluid mechanics, computer vision, and materials science
140. J.A. Sethian, Evolution, implementation, and application of level set and fast marching methods for advancing fronts. J. Comput. Phys. **169**(2), 503–555 (2001)
141. J.A. Sethian, P. Smereka, Level set methods for fluid interfaces. Ann. Rev. Fluid Mech. **35**, 341–372 (2003)
142. J. Simon, Compact sets in the space $L^p(0, T; B)$. Ann. Mat. Pura Appl. **146**, 65–96 (1987)
143. G. Son, V.K. Dhir, Numerical simulation of film boiling near critical pressures with a level set method. J. Heat Trans. **120**(1), 183–192 (1998)
144. M. Sussman, P. Smereka, S. Osher, A level set approach for computing solutions to incompressible two-phase flow. J. Comput. Phys. **114**(1), 146–159 (1994)
145. S. Tanguy, T. Ménard, A. Berlemont, A level set method for vaporizing two-phase flows. J. Comput. Phys. **221**(2), 837–853 (2007)
146. J.E. Taylor, J.W.W. Cahn, C.A. Handwerker, Geometric models of crystal growth. Acta Metall. Mater. **40**(7), 1443–1474 (1992)
147. R. Temam, *Navier-Stokes Equations*. Studies in Mathematics and its Applications, vol. 2, rev. edn. (North-Holland, Amsterdam, 1979). Theory and numerical analysis, With an appendix by F. Thomasset
148. C. Truesdell, The influence of elasticity on analysis: the classic heritage. Bull. Am Math. Soc. **9**(3), 293–310 (1983)
149. S.W.J. Welch, J. Wilson, A volume of fluid based method for fluid flows with phase change. J. Comput. Phys. **160**(2), 662–682 (2000)
150. S. Weller, Time discretization for capillary problems. Doctoral Thesis, Friedrich-Alexander-Universität Erlangen-Nürnberg (FAU) (2015)
151. S. Weller, E. Bänsch, Time discretization for capillary flow: beyond backward Euler, in *Transport Processes at Fluidic Interfaces*. Advances in Mathematical Fluid Mechanics (Birkhäuser/Springer, Cham, 2017), pp. 121–143

152. J. Wloka, *Partial Differential Equations* (Cambridge University Press, Cambridge, 1987). Translated from the German by C. B. Thomas and M. J. Thomas
153. Y.F. Yap, J.C. Chai, K.C. Toh, T.N. Wong, Y.C. Lam, Numerical modeling of unidirectional stratified flow with and without phase change. J. Int. Heat Mass Trans. **48**(3–4), 477–486 (2005)
154. E. Zeidler, *Nonlinear Functional Analysis and its Applications. I* (Springer, New York, 1986). Fixed-Point Theorems, Translated from the German by Peter R. Wadsack

Printed in the USA
CPSIA information can be obtained
at www.ICGtesting.com
LVHW080303240124
769695LV00007B/511